BEHOLD OUR GREEN MANSIONS

BEHOLD OUR GREEN MANSIONS

A Book About American Forests

WITH PHOTOGRAPHS AND MAPS

Forest Restoration · Soil Conservation · Timber Flood Control · Wild Life · Grazing · Scenic Beauty Recreation · Preservation · Use

By RICHARD H. D. BOERKER

Chapel Hill THE UNIVERSITY OF NORTH CAROLINA PRESS

*

Second Printing, JUNE 1946

To the Memory of My Father

JOHN BOERKER

Devoted Conservationist

PREFACE

LATELY OUR FORESTS have gone to war. Wood has been classified as a critical war material, and the forests of the nation are being called upon for a tremendous output of raw materials essential to the prosecution of the war. Pressure to supply this demand has inaugurated a new era of destructive cutting not only in our remaining old-growth stands but also in our promising second-growth forests of the East. Uneconomic stripping of our immature growth threatens the ability of these forests to continue to supply adequate raw materials for both local and regional wood-using industries. Unless this liquidation of essential growing stocks is halted, we shall eventually be faced with a greatly reduced output. As a result of this we shall be compelled to strike a balance between growth and consumption far below the potential capacity of our great forest areas and of the industries which depend upon forest products.

Our production of essential wood commodities for war has been handicapped by our past gross neglect of our forests. We are now reaping the distasteful fruits of decades of forest abuse. We are forced to obtain 40 per cent of our lumber from the Pacific Coast, thus increasing our manpower, equipment, and transportation shortages. Other regions have large sawmill capacity but do not have the timber to utilize it fully. We are experiencing increasing difficulty in finding high-quality timber for a number of specialty uses. Throughout the Northeast there are serious deficiencies in timber of commercial quality essential to war industries located in the region. As a result of all these inequalities, increasing requirements for lumber have been far in excess of current production.

While our forests can and will be used to the maximum to supply war needs, they must not be unnecessarily stripped of essential immature growing stock when good forestry practices can maintain the volume needed equally well. This "clear" cutting of our young forests is a serious threat to the continuity of many local forest industries and to the surrounding populations depending upon them. Serious hardships threaten these people after the war if the resources which give them a livelihood have been depleted. As a matter of fact, with our virgin timber supply rapidly coming to an end, our only hope for a continuous yield of timber for future needs lies in the growth that has sprung up since the original forests have been cut. The growing stock on these forest lands, instead of being further depleted, must be restored and rehabilitated to somewhere near their potential capacity to produce.

One may well ask, "Why is forest restoration necessary to present national defense and post-war human welfare?" or, "What do forests contribute both to war and to peace?"

Starting with actual warfare itself, the daily papers remind us that forests are useful for hiding troop movements and concentrations, gun implacements, tanks, and other mechanized equipment. They also serve the extremely important function of hiding military airplanes while on the ground.

Wood has thousands of uses for national defense. It is estimated that about two billion feet of lumber are needed to house each million men for the army. This is exclusive of the millions of feet needed to house the hundreds of thousands of defense workers. Nor does this take into account the millions of feet used in those defense industries in which wood is taking the place of steel. Many millions of feet also go into the ships of our navy and our merchant fleet. Even steel ships are supported during construction in cradles made of wood. Our air force requires laminated wood propellers, struts, and spars. Pontoons used both by men and by mechanized equipment to cross rivers are also made of wood. In the shipment of war materials to our allies, the crates, boxes, barrels, and other containers are made of wood or wood products. And so the story goes. It is difficult to see how the war could be successfully carried on without the use of wood or its many by-products.

In our defense effort wood is of prime importance to maintain the transportation and communication systems. Freight cars, railroad ties, bridges, telephone and telegraph poles, wharves, docks, docking, and mine props and supports are mostly made of wood.

When metals are restricted to military needs, wood or one of its numerous products or derivatives, is one of the first materials to which manufacturers turn for a substitute.

Other forest products are useful in building and maintaining defense industries. This includes all the well-known building materials made from wood. Paper, paper products, cellophane, and other plastics are made of wood cellulose. The farmer needs fuelwood, fence posts, wood for silos, veneer for fruit boxes, and a host of other wooden products to produce food for defense. Cork provides linoleum for battleships, lifebelts, carburetor floats, and engine gaskets.

The chemical products made by the distillation of both hardwoods and softwoods, from naval stores, and from wood cellulose itself, form an almost endless array of chemicals, solvents, explosives, and plastics, many of which are of the utmost importance for the operation of the army, navy, and air forces.

The distillation of hardwoods like beech, birch, and maple yields acetate of lime, wood alcohol, and charcoal. The acetate is valuable for making acetone (an important solvent), smokeless powders, cordite, and other high explosives. It also finds use in the leather and textile industries. Wood alcohol is a solvent used in the paint and varnish industries, in making aniline dyes, and in the manufacture of formaldehyde, photographic films, smokeless powders, and nitro-cellulose explosives. Charcoal is valuable for making high grade steel, for making gunpowder, and for refining gasoline.

The distillation of pine wood furnishes turpentine, pine oils, tar oils, tar, and charcoal. Tar oils are used to make disinfectants, creosote, and paint driers; while tar is an essential in the building industries. Turpentine is of immense importance to our navy because it is the solvent of most paints used. It is also important in the production of synthetic camphor, which in turn is essential in the manufacture of smokeless powders.

A research chemist of the Hercules Powder Company recently discovered that pine oil is an efficient separating agent in the flotation process by which some metals are freed of their impurities. Also a new resin product has been found valuable in shortening and facilitating the production of Portland cement and also as an emulsifier and stabilizer of asphalt. Thus, these forest by-products contribute to the quicker and cheaper construction of highways, airfield runways, gun implacements, and other structures used in military defense and in any phase of war industry in which cement, concrete, or asphalt is employed.

As for the use of wood and forests in the post-war world, very little could be added to the statement found in the annual report of the Chief of the Forest Service for 1942:

"The United States, and the world will need wood for the post-war reconstruction. Wood will be needed for our long-term development and progress as a democratic nation. Abundant natural resources are essential to the realization of our ideal of peace and liberty; they can be maintained only through farsighted conservation. We do not want to find, after the war is over, that we ourselves have unnecessarily impaired or destroyed one of the basic resources of the land we have been fighting to protect. We cannot afford to temporize when adverse results will be so far reaching and so long sustained as those from needless destruction of productive forests."

Indispensable as lumber and other forest products are in our national defense, they are no more important than certain other services rendered by the nation's forests. Recreation upholds national morale. Forest recreation—camping, fishing, hunting, etc.—helps to relieve the strain of long hours of work.

Food is an absolute essential both for marching armies and for civilian workers, not to mention the feeding of starving peoples whose farm lands and homes have been ravaged by war. Any factor, therefore, that safeguards productive farms, is vital to defense. Forests help to prevent erosion and floods, two of the greatest enemies of farm lands. The forest is also the source of a large part of the forage used to feed livestock. This stock means meat, leather, wool, and other products for defense. Finally, the farmer is deeply indebted to forest bird life for helping to destroy injurious insects which prey on vital food crops.

The control of erosion and floods is a vital necessity for national defense since these destructive agencies often destroy military roads, railroads, bridges, airfields, and other communication and transportation facilities which may be of great strategic value. They also destroy reservoirs which store water.

Essential hydro-electric power for war industries is assured only if the streams which furnish it are protected from floods and the mountain sides from erosion. Only the forest-covered watershed can furnish a uniform supply of silt-free water.

The domestic water supply of cities, towns, villages, as well as of army cantonments must be safeguarded from pollution. A steady supply of pure, fresh water, free of surface wash, is as important as food itself. Forests regulate and protect this supply. That is why some municipalities, many years ago, began to reforest their water-supply lands.

Finally, after the war emergency is over, the forest will no doubt play an important role in employing labor until non-war industries recuperate. Federal relief agencies have already done some work of this kind, but much remains to be done. Fire protection, reforestation, the building of forest roads, trails, and structures, the control of insects and diseases, thinning and improvement cuttings, and wild-life management will give employment to help cushion the shock of changing from a wartime to a peacetime economy. Forest restoration on a grand scale, greater than that ever attempted, might well be an important contribution to the recovery of national economic stability.

Thus in wartime and in peacetime forests offer employment for millions not only in getting the raw materials out of the forest, but also in converting these materials into thousands of commodities useful to our civilization. Hence, forestry, practiced on a wider and more universal scale to insure a sustained annual yield of valuable products and services, must always be an important part of the household economy of civilized nations. As we shall see later, private, federal, state, and local agencies must bear the burden of this formidable task.

Finally, the ultimate objective of all forest management must be human welfare, for forestry is part of the larger problem, close to every one of us, of making America a more beautiful place in which to live—from woodland vistas and clear trout streams, to healthy, happy, and busy forest communities worthy of a great democracy. And this is the challenging problem of forest restoration.

ACKNOWLEDGMENT

A LARGE NUMBER OF PERSONS have contributed helpful criticisms and suggestions in the writing of this volume. I am especially indebted to staff members of the Division of Information and Education, U.S. Forest Service, for checking the data; and to officials of various lumber associations for their contributions to Chapter XVIII.

The following government agencies and private corporations have contributed useful material in the form of photographs or printed matter: United States Forest Service, Soil Conservation Service, Canadian Travel Bureau, Maine Development Commission, New York State Conservation Department, Pennsylvania Department of Forests and Waters, New Hampshire State Forester, Society for the Protection of New Hampshire Forests, New Hampshire Aerial Tramway Commission, Forest Products Laboratory (Madison, Wisconsin), E. I. du Pont de Nemours & Company, Weyerhaeuser Timber Company, West Coast Lumbermen's Association, American Forest Products Industries, Inc. (Washington, D. C.), Southern Pine Association.

As far as possible I have availed myself of original sources of material for this book. I must acknowledge the extended use of *A National Plan for American Forestry* (Senate Document No. 12, 1933) and of *The Forest Lands of the United States* (Report of the Joint Committee on Forestry, 1941). Scores of other federal, state, and private documents have been consulted, and these are cited in the text. Also considerable unpublished information from federal, state, and private sources has been used.

I must acknowledge also the careful work and good advice of the editorial staff of the University of North Carolina Press, and especially the deep personal interest of Mr. Wendell Brooks Phillips, now of the National Archives in Washington, D. C., whose valuable editorial comment and out-of-doors knowledge of conservation greatly aided in achieving the final draft of this volume.

Finally, I am indebted to my daughters Huldah and Ruth for help in typing, the latter having accomplished alone the original draft of the manuscript; and to my wife for much careful help in proofreading.

In this book I have tried to present the facts of American forestry—and controversial issues have been largely avoided in order to emphasize more strongly those fundamental principles on which foresters are generally agreed. For the opinions expressed and for any errors that may have occurred, the author alone is responsible.

RICHARD H. D. BOERKER

Kingston, New York
June 1, 1944

CONTENTS

BEHOLD OUR GREEN MANSIONS

CHAPTER I

WHAT IS CONSERVATION?

THE CAMPER who fails to put out his fire or the careless traveler who flips a match from his passing car, generally fails to visualize the possible results of his carelessness. Years ago, as I was camped beneath the tall giants of a western fir forest, the tragedy of forest fires was branded into my soul.

For several months I had been in the woods, and spent many an evening watching the sunbeams filter through the branches and stately trunks, much as light streams through the stained-glass windows of a cathedral. Sometimes after sunset I would walk down to the lake and there from a rocky promontory gaze into a star-studded sky pierced by the black spires of treetops on the opposite shore. The air was cool and fragrant. The only sounds were an occasional splash of a trout on the lake, the whirr of wind through fir branches, and the persistent hooting of an owl. Finally relaxed, I would roll up in a blanket and go to sleep.

On some evenings, however, too restless for continuous sleep, I would awaken and revisit my shrine—that moonlit lake whose languid waters lapped at the roots of the century-old firs that lined the shore. There I could review the happenings of the day and plan for tomorrow, or simply contemplate the glimmering lake. Though solitary, I had no sense of being alone. The whole forest world was around me—and the occasional clank of hobble-chains assured me that my faithful mare was not far away.

A few days later, tragedy came. Some neglected campfire, driven by a strong dry wind, transformed this forest paradise—a playground of bounding deer and gamey trout, a haven of cool, deep, and primitive verdure—into a roaring inferno which would have defied even Dante's power of description.

The roar of a forest fire has been aptly compared to "a thousand freight trains crossing steel trestles." True enough—but the monotonous roar was only an undertone or background to the near-by hisses, shots, and blasts. There were rifle-like explosions as the flames leaped high through the treetops. Wooded canyons were transformed into colossal chimneys, through which the flames ascended with the speed and intensity of a blast furnace. Finally a tornado-like whirlwind swept into the center of the inferno, uprooting trees that blazed like giant torches and scrambling them together with a fearful rending sound as thousands of burning embers shot high into the air. No living creature could outrun that withering blast, or survive it. Every sort of animal and bird was charred to a crisp. Even the fire fighters were forced to keep fully a mile away from the advancing flames.

Each night as the wind died down, fire-fighting crews would push their work at top speed. Occasionally they would pause to watch the red glow of the sky or listen to the crash of some huge dead snag that toppled down with a shower of sparks. Our eyes burned like balls of fire and our lungs were filled with suffocating smoke and the sweetish odor of resinous fumes. Finally, with body and brain exhausted and with spirit almost conquered, we wrapped up in our saddle blankets and stretched out on a bed of cooling ashes. For three days and three nights we endured this soul-sickening experience, until at last a drenching rain put an end to the tragedy. I knew I was lucky to be alive.

Shortly afterward, I made a trip over the scene of destruction. Bubbling mountain brooks had shriveled to dust. Trout floated belly up in the ash-covered lake. Not a living thing was in sight—not even a buzzard.

Here I found the antlers of a once-proud buck; over yonder the bloated corpses of a half-dozen does and fawns who had scurried into the vain protection of a mountain ravine; still farther, the tangled wreckage of a once-thrifty sawmill village. On every side, as far as the eye could see, there was a graveyard of charred tree-skeletons, with heaps of gray ashes and a spell of unforgettable desolation over all. To paraphrase the words of H. L. Mencken, "It seemed incredible that mere ignorance should have achieved such a masterpiece of horror."

As I contemplated the results of one individual's criminal carelessness, I asked myself: "What does it all mean? We prate about all-powerful kings and dictators. Who has turned this irresponsible fool loose in the woods, to turn heaven into hell and blacken the path of generation after generation of law-abiding citizens?" Certainly here was concrete evidence to support G. Stanley Hall's observation that "man has not yet demonstrated that he can remain permanently civilized." Thoreau would no doubt have complained ironically that we had "civilized our forests off the face of the earth" —or perhaps Santayana was right when he said that "Wisdom comes by disillusionment."

When it was all over, this holocaust went into the records as a mere statistic—location, area burned over, amount of damage done, cause of the fire, and "remarks." That was the end of the story. Yet in an average year, about 40,000,000 acres of American forests go up in smoke. Most of the fires are man-made, and at least a quarter of them are intentionally set. The hideous mutilation and disfigurement of our landscape is, of course, only part of the picture. There is also the loss of lumber and of employment to lumber workers—and even more serious, the lasting damage to forest resources, soil, and watersheds.

Fortunately, there have been men in America who took the lead in saving our great heritage from destruction: scholars like Charles R. Van Hise, naturalists like John Muir, public administrators like Theodore Roosevelt, foresters like Gifford Pinchot. These are great names, and the movement which they initiated has become a part of our national policy. Yet the public has often misunderstood the aims of conservation: the word itself has been taken over by sentimentalists, who connect it with such superficial notions as the building of bird houses or "planting a tree for every one cut down." Many people even think of conservation as a complete withdrawal of forests and other natural resources from use. Nothing could be further from the true purpose of conservation, which was brilliantly expressed by John Ruskin more than sixty years ago:

> God has lent us the earth for our life. It is a great entail. It belongs as much to those who are to come after us as to us, and we have no right, by anything we do or neglect, to involve them in any unnecessary penalties, or to deprive them of the benefit which was in our power to bequeath.

Here we get the idea of the earth as a great natural heritage which we should wisely use for our own benefit but pass on to our descendants unimpaired. Such an ideal is wholly practicable. Under the right sort of management, soils can produce good crops and yet continue to grow steadily richer; properly handled, forests can produce a permanent supply of lumber and yet maintain themselves with new growth which replaces mature trees that are cut. This is the true conception of conservation, and it is the direct opposite of non-use or narrowly restricted use.

Thus interpreted, the principle of conservation is the guiding theme of this book—but since the word "conservation" has been so misunderstood and misused that it has taken on a certain ill repute, it has been discarded in favor of more up-to-date terms that will run no danger of being misinterpreted. Our whole emphasis will be on *forest restoration for multiple use,* with human welfare as the ultimate object. We must stop forest fires and keep America green—but we must also utilize our forests for the maximum benefit to ourselves and to those who come after us.

CHAPTER II

FOREST RESOURCES OF THE UNITED STATES

IN THE EARLY DAYS of American settlement, the forest provided many of the necessities of life and furnished a large part of the most vital material for building towns, cities, and industries. An abundant and accessible supply of softwood timber was quickly converted into houses, gristmills, schools, and churches. Near-by woodlands furnished logs to burn in the fireplaces, and wood of oak, maple, and walnut for the construction of furniture. Ash, elm, and hickory were readily available for making wagons, plows, and handles for hoes and axes. Virgin soil, enriched by centuries of forest growth, brought forth abundant crops. Fish and game of the forest gave variety to the family larder, and fur-bearing animals supplied an early source of income for the pioneer. New regions were explored with canoe and batteau, and soon wooden ships plied the rivers and lakes. Ocean-going commerce later carried our products to the four corners of the world. Railroads spanned the country on wooden ties, and telephone and telegraph wires were strung on wooden poles. Newspapers made of wood cellulose carried the news to the far-flung corners of the American empire. Little by little these transportation and communication systems knit the states into a united nation.

More recently forest products have helped to build the great industries which have turned out our airplanes, automobiles, radios, modern household appliances, and many other present-day necessities. Today our forests are going to war at an ever-increasing rate, furnishing materials without which victory could not be achieved.

Even we who love the forest do not deplore the billions of feet of pine, spruce, and oak that have been cut to make America great. Our only regret is that so much timber has been wasted and burned, and that our forests have been exploited without due regard for the future needs of the nation. We have acted like prodigal sons with no thought for tomorrow. We have slashed and burned, apparently believing these vast resources to be inexhaustible. The twentieth cen-

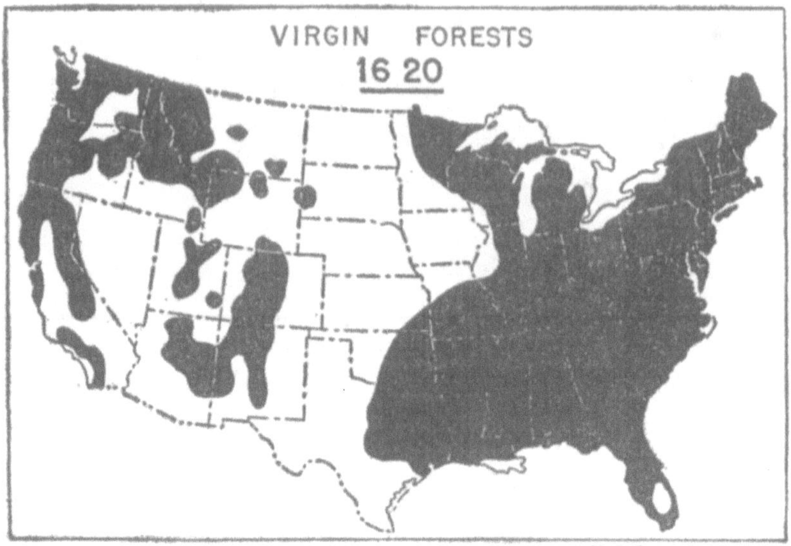

U. S. Forest Service Photo

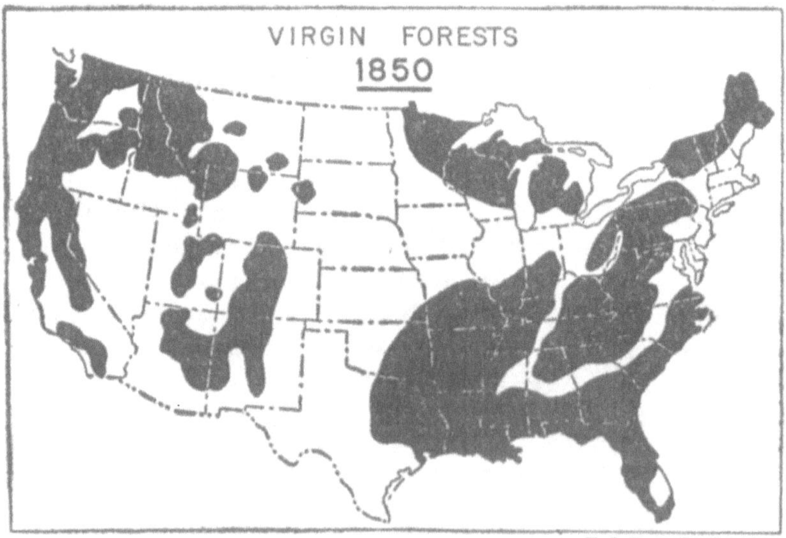

U. S. Forest Service Photo

tury finds us now with our virgin forests nearly gone and with a tremendous demand for wood both here and abroad looming up after the war, in the face of a rapidly dwindling world supply.

The first national timber reserves were set aside for posterity about fifty years ago. During this time much conservation work has been done. We can look with justifiable pride upon our record of

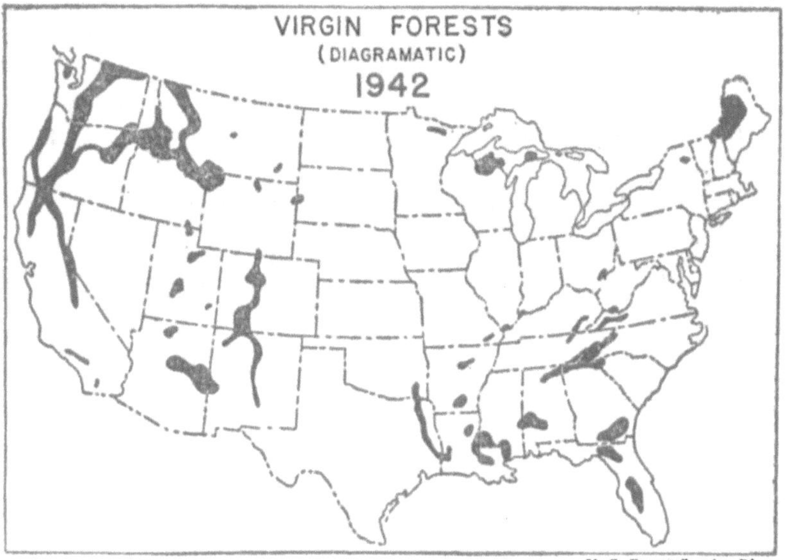

U. S. Forest Service Photo

achievement in bringing 30 per cent of our total forest area under public control. These national, state, and community forests assure at least some timber for the future. Our great failure is in our lack of a coördinated and sustained policy towards the major portion of our forest resources—the 70 per cent still held in private ownership.

Too many people have the idea that our public forests solve the conservation problem and that from now on the movement will be carried forward under its own momentum. Nothing could be further from the truth. Too many public agencies are handling the conservation problems. There is too much conflict of purpose. Unified control and a common objective are sadly needed. There is always the danger that good forestry may be wrecked by bad politics. Furthermore, the public conscience has not yet awakened to the fact that a sound American policy for private timberland management has not been formulated. Progress has been slow and difficult, and public support must be developed. No forest conservation program can be complete without a plan for perpetuating growth on the 70 per cent of our forest area that comprises some of the best timberland we have.

Within the public forests and in many of the private ones, fire protection has made much progress. But when we learn that between thirty and forty million acres of woodlands are still being burned over annually, and that over 90 per cent of these fires are set by human carelessness or design, we must conclude that complete forest fire protection is far from an accomplished fact. In our public forests we are providing a future crop; in most of the private forests we are not. Some large private timber holders are establishing "tree farms," but, on the whole, forest restoration on private lands is painfully slow.

All this is on the utilitarian side. We are still far from developing a passion to preserve the beauty and the majesty of our American forests. We are still too much concerned with what dollars and cents they will bring in, and not enough interested in making this a more beautiful country in which to live. Our quest for beauty does not seem to go beyond our own doorsteps. Fire, erosion, and irresponsible methods of logging are still converting beautiful hillsides into hideous blots on the landscape.

As individuals we honor and respect the father and mother who gave us birth, and who guided and taught us through our helpless years so that we might become good citizens. As a nation we have not yet learned to honor and respect our great forests, which have supplied such a large part of our food, fuel, shelter, and raw materials. We are not yet forest-conscious; nor have we yet made forest culture, with its vast social and economic benefits, a part of our religion—an eleventh "Commandment." Public education has missed a great opportunity. No broad and sound forest policy can be born until there is more and better understanding of our great forest empire and its many problems. Our major educational channels must provide the groundwork.

In order that the reader may appreciate the vast extent and the wide distribution of our forest resources, their great diversity both as to quantity and quality, and the enormous economic, political, and financial problems involved in their ownership, it will be necessary to describe briefly our forest areas. Since orderly, continuous forest management is the ultimate objective of all sound private and public forest policies, we shall consider the distribution of our most important tree species and enumerate the important types of management which are now being used in basic methods of reproduction. An appreciation of our forest resources will help to make us better citizens, aware of the big conservation job to be done, and more alert and aggressive in carrying out the measures which will insure their perpetuation.

FOREST AREAS

From Ponce de Leon and De Soto through the centuries to Lewis and Clark, Pike, and Frémont, explorers and settlers alike gazed in awe and amazement at the unbroken forests that stretched to the horizon in every direction. Our history and literature are full of references to the extent of our forests and to the size of some of the individual trees. In Maine, Thoreau stood in reverence before a six-foot giant white pine; in Southern Illinois, Audubon built his campfire beside a six-foot white ash log; on the Pacific Coast, Douglas marvelled at the ten- and twelve-foot fir trees which now bear his name; and in California, giant Sequoias twenty-seven feet in diameter and four to five thousand years old, are classed with the other great wonders of the world. Our original forests covered almost without a break all the land from the Atlantic seacoast to the Mississippi River. Millions of additional acres of woodland stretched over the middle and upper slopes of the Rockies, the Sierras, and the Cascades. On the humid Pacific Coast was found one of the heaviest stands of timber anywhere in the world; here were single trees that would yield enough lumber to build ten one-family cottages. Truly America was endowed with a rich heritage.

The original forest is said to have covered about 820,000,000 acres, or about 43 per cent of the total land area of the continental United States. Three centuries of agricultural and industrial development have reduced this to 630,000,000 acres, or about 33 per cent of the total land area. This is thought to be the minimum area that will be needed to grow the wood for our industries and for export, besides providing the numerous other services rendered by the forest.

Area alone, however, does not give a true picture of the present timber situation, because some acres of forest bear a much heavier stand of timber than others. Of the original 5,200,000 million board feet of timber that is estimated to have stood in this country three hundred years ago, only about one-third, or 1,763,651 million board feet, are left standing, according to 1938 estimates. Most of the cutting and destruction of our forests has occurred in the eastern half of the country; hence most of the remaining stands are in the West, two or three thousand miles from the populated centers where they are needed.

Timber Regions

The United States has been divided into eight regions, based upon the location of the present commercial saw timber (map, p. 11). These regions do not necessarily coincide with the natural geographic regions outlined later in this chapter.

The Northeastern Region comprises the New England states, New York, New Jersey, Pennsylvania, Delaware, and Maryland. Its principal commercial species are birch, beech, maple, spruce, fir, white and red pine, hemlock, and oak. It contains only 5 per cent of the standing timber of the nation. Since this is the most populated section of the country its annual timber needs are very great. Most of what it uses must be brought in from other states.

The Central Region covers the states of Iowa, Missouri, Illinois, Indiana, and Ohio, and it is an important source of hardwoods such as oak, birch, beech, maple, red gum, yellow poplar, hickory, and a small number of softwoods such as southern yellow pine and cypress. The total present stand is small and the region is valuable chiefly for agricultural products. It has only 1 per cent of our present stand.

The Lake Region includes the states of Minnesota, Wisconsin, and Michigan, which once had an abundant stand of red and white pines. The main stand today is of birch, beech, maple, and hemlock, with smaller amounts of spruce, fir, red and white pine, and oak. Only 3 per cent of our total stand is located in this region. With a cold climate and sandy soils, this is predominantly a forest-producing region.

The Southern Region comprises the fourteen states found south of the three regions just enumerated. It contains 44 per cent of our commercial forest area and 22 per cent of the total saw-timber stand, and is, therefore, one of our most important forest regions. The South has just begun to realize the place it must play in the conservation of forests. With its warm climate and predominantly sandy and gravelly forest soils, it has an enormous potential capacity for timber production. In the southern region an important timber tree is the southern yellow pine, though other softwoods such as spruce, fir, hemlock, and cypress are also found. In many parts of the South, hardwoods are highly important, since about 65 per cent of the total remaining volume of saw timber of these precious species are in this region. They reach their best development in the moist, rich bottomlands of the region. Oaks, red gum, yellow poplar, tupelo gum, beech, maple, and many other species make up the stand.

The Plains Region is mostly treeless except for a fringe of growth along the streams. A considerable amount of western yellow pine in the Black Hills constitutes the largest stand of commercial timber available. The region embraces the tier of states from North Dakota to Texas.

The Columbia River Basin includes the states of Washington, Oregon, Idaho, and part of Montana. This comprises only 16 per cent of our commercial forest area, but 50 per cent of our total saw-timber stand. Its importance as a source of future timber can hardly be over-estimated. Its principal species are Douglas fir, western yellow pine (ponderosa), western hemlock, true firs, spruce, western larch, western white pine, sugar pine, and several others. The moist, mild climate and predominantly volcanic soils are very favorable for timber production.

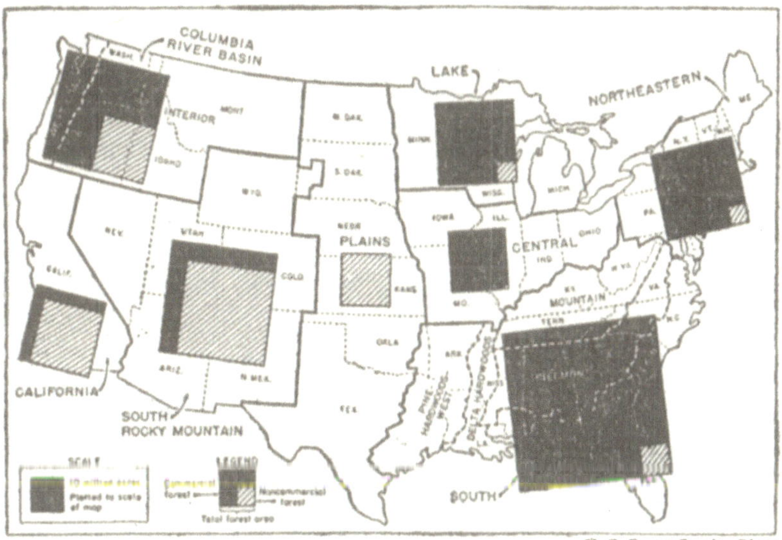

U. S. Forest Service Photo

RELATIVE FOREST AREAS, BY BROAD CLASSES AND REGIONS

The California Region, which has only 3 per cent of our commercial forest area, contains 12 per cent of the total stand of saw timber. The species are mainly western yellow pine, Douglas fir, true firs, redwood, and sugar pine. A mild climate and a mountainous terrain combine to make most of the area valuable chiefly for timber and livestock. Only the large inland river valleys are agricultural.

The Southern Rocky Mountain Region comprises the states of Wyoming, Nevada, Utah, Colorado, Arizona, and New Mexico. This region has 7 per cent of our commercial forest area and 7 per cent of the total saw-timber stand. The species are western yellow pine, spruce, lodgepole pine, Douglas fir, and the true firs, with a scattering of others. Most of the timber is at high elevations and to a great extent inaccessible according to present logging standards. The valleys, where irrigated, produce agricultural crops and livestock. While the timber may be utilized in the distant future, at present the forests are of most vital importance in protecting the watersheds used for irrigation.

Table 1 *

OWNERSHIP OF FOREST LAND IN THE CONTINENTAL UNITED STATES EXCLUSIVE OF ALASKA, 1938

Class of Ownership	Per Cent of Grand Total	Total Area	Commercial Area	Non-Commercial Area
		Millions of acres		
Private				
Farm woodland	29	185.5	138.8	46.7
Industrial and other	40	248.3	202.1	46.2
Total	69	433.8	340.9	92.9
Public				
Community	1	7.8	7.1	.7
State	3	19.0	16.9	2.1
Indian Reservations	2	12.0	6.4	5.6
National parks and monuments	1	6.5	. . .	6.5
Public domains	4	24.0	4.7	19.3
National Forests	19	122.0	81.5	40.5
Other federal	1	5.0	4.2	.8
Total	31	196.3	120.8	75.5
Grand Total, all classes	100	630.1	461.7	168.4

* Source: *Forest Lands of the United States,* House Report 323, 1941.

Table 2 *

OWNERSHIP OF STANDS OF SAW TIMBER IN THE UNITED STATES BY REGIONS, 1938

Regions	Per Cent of Total	Total Stand	Publicly Owned or Managed	Private
		Millions of board feet		
Northeastern	5	84,025	2,950	81,075
Central	1	14,301	531	13,770
Lake	3	57,616	10,917	46,699
South	22	386,570	16,501	370,069
Columbia River Basin	50	882,632	500,138	382,494
California	12	213,480	89,121	124,359
South Rocky Mountain	7	124,992	116,761	8,231
Plains	0	35	35	.....
Total	100	1,763,651	736,954	1,026,697
Per cent of Total		100	41.8	58.2

* Source: *Forest Lands of the United States,* House Report 323, 1941.

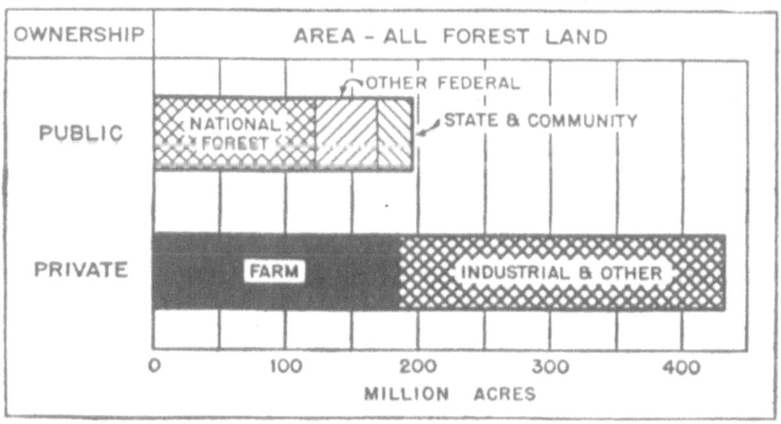

U. S. Forest Service Photo

OWNERSHIP OF FOREST LAND

Only 73 per cent of our forest area is today considered valuable for commercial uses; the rest, because of remoteness, rough topography, or species of low value, does not qualify for economic timber growing, although it may be valuable for watershed protection, recreation, and other uses. Of our total forest area, 69 per cent is privately owned and 31 per cent is publicly owned. The private holdings have, in general, much better stands of timber, as is shown by the fact that 78 per cent of all private area is classified as "commercial," while only 61 per cent of the publicly held forest land is so classified. Also of much importance is the fact that 29 per cent of all our forest land consists of farm woodlands, which indicates the relative importance of these lands to timber conservation. About 19 per cent of our total forest land is in the National Forests and 40 per cent is in large industrial holdings.

Table 2 shows that 58.2 per cent of our present stands of timber (as distinguished from forest area) are in private hands as compared to 41.8 in public ownership. So both in point of area and of volume, privately owned forests are greater than publicly owned ones. However, there is this important difference: most of the private forests are being handled without any regard for future needs while all of the public stands are either reserved from cutting or handled on a sustained-yield basis. Consequently, as time goes on, public forests will be of much greater significance.

The location of saw-timber stands is also significant, as three regions—the South, the Columbia River Basin, and California—contain 84 per cent of our total saw timber. Percentages of public ownership are distributed as follows: 93 per cent of the saw timber is publicly owned in the southern Rocky Mountain region, 76 per cent in the Columbia River Basin, and 40 per cent in California. Public ownership of timber stands reaches a very low point in the South, where only 4 per cent is publicly owned. Considering the huge area involved, there is great need for more National and State forests in the southern states.

Character of Commercial Stands

Of the 462,000,000 acres of commercial timberland only 46 per cent is covered with saw timber. Cordwood stands make up 22 per cent; satisfactory restocking stands, 15 per cent; and stands which are poorly restocking, or not at all, make up the remaining 17 per cent. The latter figure is important in that it indicates that artificial restocking will be necessary in many of these areas if they are ever to produce a stand of timber.

Table 3 shows the distribution of commercial forest areas by regions and by type of growth. The South has the greatest acreage of forest land fit for saw timber, but it also has large areas—14 per cent of the total—where restocking with new growth is poor or entirely lacking. In the Lake states there is over 40 per cent of this type of land, or 21,000,000 acres, in need of reforestation if they are to produce stands of timber within a reasonable time. The financial problem involved is probably greater than the states could handle effectively.

Table 3 *

COMMERCIAL FOREST AREAS BY REGIONS AND KIND OF GROWTH, 1938

Region	Per cent of Total	Total Area	Saw-timber Area	Cord-wood Area	Fair to satisfactory re-stocking	Poor to no re-stocking
		Millions of acres				
Northeastern	13	59	21	15	15	8
Central	6	29	10	9	5	6
Lake	11	52	7	11	13	21
South	44	203	97	48	29	29
Columbia R. basin	16	74	44	12	9	9
California	3	14	11	†	†	2
Southern Rocky Mountains	7	31	23	6	†	2
Plains states	0	†	†	†	†	†
TOTAL	100	462	213	101	71	77

* Source: *Forest Lands of the United States*, House Report 323, 1941.

† In rounding off the figures to the nearest million these areas were too small to appear in this table.

IMPORTANT COMMERCIAL SPECIES

So far as the present supply of timber is concerned, the eastern states contain 31 per cent and the western 69 per cent. In the East this is about evenly divided between hardwoods and softwoods. In the West practically all the standing timber is of the softwood variety. Taking the country as a whole, 84.6 per cent of our total supply is of softwoods and only 15.4 per cent of hardwoods.

There are between sixty and seventy important commercial species of wood on the lumber market today. The most abundant species standing is Douglas fir, comprising 27.8 per cent of the total supply. Moreover, three species—Douglas fir, western yellow or ponderosa pine, and the southern yellow pine—make up 51.7 per cent of our total supply. In fact, eighteen species or groups of species (grouped because their woods are similar) make up over 90 per cent of our total supply. This indicates that as the less abundant species become exhausted we will have to rely on these familiar staple species to take their place. (See Table 4.)

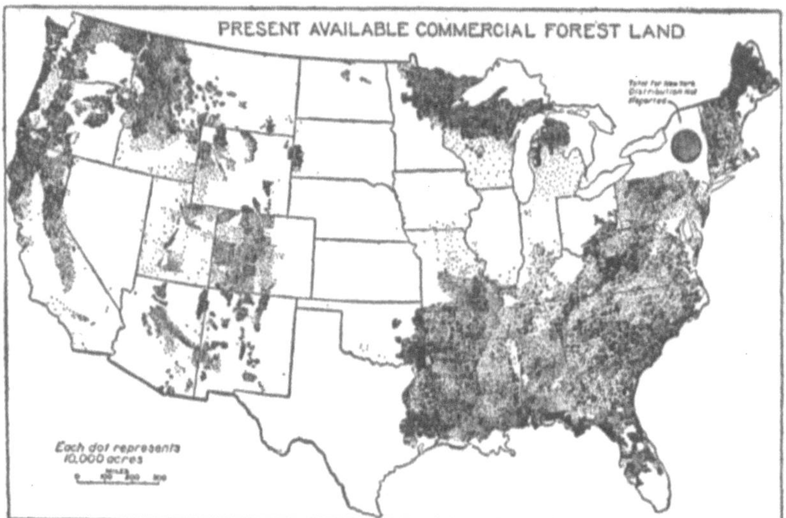

U. S. Forest Service Photo

Diversity of Present Conditions *

Let us turn now from the quantitative aspect of our forest to a qualitative view, and from a consideration of their present importance to a forecasting of their future development.

As we are still cutting timber faster than it is growing, we are still using up our wood capital and mortgaging the future. In order to balance the annual growth with the annual cut, the best practices of scientific forest management must be instituted. Each forest type or tree-combination must be given the treatment required to make it most productive, whether the objective be timber, pulpwood, watershed protection, livestock management, wild life, or recreation.

* This section, as well as the description of forest regions and types, is based upon scores of government and state bulletins from all parts of the country, together with personal study and observations made by the author in every state in the Union.

The character of this treatment on any given area will depend largely upon the physical characteristics of the trees that make up the forest.

Table 4 *

STAND OF SAW TIMBER IN THE UNITED STATES BY SPECIES, 1938

SPECIES	MILLIONS BOARD FEET	PER CENT OF TOTAL
Douglas fir	489,905	27.8
Ponderosa pine	224,904	12.7
Southern yellow pine	196,840	11.2
True firs (West)	121,737	6.9
Western hemlock	115,551	6.5
Oak	83,700	4.7
Western spruce	62,821	3.5
Birch, beech, and maple	55,178	3.1
Redwood	39,150	2.2
Lodgepole pine	38,620	2.1
Red gum	27,837	1.6
Western larch	25,306	1.4
Sugar pine	24,684	1.4
Eastern spruce and fir	24,086	1.4
Tupelo gum	20,403	1.1
Western white pine	18,333	1.0
Eastern white and red pines	18,101	1.0
Eastern hemlock	17,552	1.0
All others	158,943	9.4
TOTAL	1,763,651	100.0

* Source: *Forest Lands of the United States,* House Report 323, 1941.

Forests are not mines of wood; nor are they storehouses of standing logs. Forests are complex communities of living and growing organisms. Each organism—each tree species—has its own behavior pattern. Each kind of tree, be it pine, spruce, cypress, oak, or maple, thrives in its own peculiar set of conditions—as to soil, light, heat, and water. When several kinds are growing together, as in the birch-beech-maple-hemlock forests of the North, the behavior pattern becomes very complex.

Each tree species and each forest type or association passes from germination to maturity through a definite life history. The forester must know these life histories and adapt his forest management to them. Upon this knowledge depends his success in handling the species in the nursery, in planting it in the field, in protecting it from its enemies, in securing its natural reproduction, and in harvesting the final crop. The forester needs accurate information on at least 100 forest types or tree combinations and on about 150 commercial forest trees. Since forestry is such a long-time business, any mistakes made in the wrong selection of species or of methods of handling them may entail large financial loss, to say nothing of years of valuable time.

The complexity of our forest types and forest management is closely linked with the great diversity of our soil and climate. We have practically every kind of soil from the deepest clays to the most shallow gravels; from dry sands to water-logged swamps; and from tidal marshes and over-flow bottomlands to high alpine tundras. Our climates vary from the sub-tropical to the sub-arctic, from the aridity of deserts with less than 5 inches of rainfall a year to the wetness of forests that are drenched with over 120. We have some of the best forest-producing lands in the world, and we also have our share of the poorest.

All this seems simple, however, beside the fact that the forester is also confronted with a complex array of man-made conditions. Untrammelled nature will ordinarily reproduce its forests in a systematic way with the species best suited to the locality. Only the species most fitted to the local conditions will survive in the long run. But man has entered the picture with axe and fire, which are followed by insects and tree diseases. To say that he has disturbed the natural succession of species and types is putting it mildly. The heavy cutting of white pine and spruce in the East has left part of the forest in possession of hardwoods of the less desirable kinds. The slaughtering of white pine in Michigan has surrendered the sandy barrens to scrub oak and jack pine. In California, valuable pine and spruce forests have given way to useless chaparral. And almost everywhere the cutting of valuable sunlight-needing species has surrendered the land to less valuable shade-enduring ones, because the former could not thrive under the latter.

Forest fires usually follow lumbering. Fires superimpose another set of problems over the old set. They usually change the physical, chemical, and biological conditions of the soil to such a vast extent that often an entirely different and inferior ground-cover springs up in place of the old one that was cut and burned. Furthermore, it may take a generation to get the soil back finally into the proper condition to grow the crop best suited to the locality. Thus in portions of New England a severe fire is followed successively by bare soil, herbs, cedar, birch and poplar, pine, oak and hickory, and finally the beech-maple-hemlock climax type of the region.

When we recall that about seven-eighths of our original forests have been cut over, and much of it also burned over, we can get a vague idea of the great variety of facts that must be known before it will be possible to grow the best crop of the most valuable species in the shortest time on each acre of forest land.

Geographical Forest Regions

A brief description of our important natural forest regions should serve a manifold purpose. It should, first of all, give the forester a basis for silvicultural management; it should also give the lover of the woods a better conception of the vastness and diversity of our heritage; the hunter, fisherman, and naturalist a closer acquaintance with the haunts of plant, animal, and bird; and the traveler a familiarity with the forest regions and belts through which he passes. It should give us all a deep sense of pride that we possess such endless variety of forest wealth, and it should arouse a greater sense of responsibility in seeing that this wealth is protected and efficiently used.

The Northern Forest occupies most of the northern two-thirds of New England, New York, Michigan, Wisconsin, and Minnesota, together with a long, narrow tongue that extends southwestward through the mountains of central Pennsylvania to northern Georgia. Its typical species are the red, white, and black spruces, balsam fir, red pine, white pine, jack pine, pitch pine, northern white cedar, tamarack, hemlock, birches, beech, poplars, basswood, elms, and maples. These species are grouped more or less distinctly into four distinct sub-regions: the northern spruce, the northern pine, the northern hardwood, and the region of the higher Appalachians.

The spruce forests grow mostly in northern New England and the Adirondack mountains of New York, although there are small areas also in the northern Lake states. The spruce is sometimes found pure and sometimes in mixture with hardwoods and with balsam fir.

The pine forests are found mostly in the northern parts of Michigan, Wisconsin, and Minnesota. White and red pines occur pure (except for some jack pine) in large areas, but are mixed with spruces and firs in the northern part of their range and with hemlock and hardwoods in the southern part.

Directly south of these two regions is a comparatively narrow belt of hardwoods mixed with hemlock and pine extending from central Minnesota across central Michigan to central Massachusetts. These regions more or less overlap at their common borders.

Lastly, the northern forest is extended southward by virtue of the high mountains of the Appalachian system, on whose upper slopes climatic conditions are similar to those found at lower elevations farther north. It is an exceedingly complex forest numbering over 135 species of trees. In the southern part of these mountains, typical northern species such as hemlock, red and black spruces, white pine, yellow birch, pitch pine, beech, sugar maple, red maple, and basswood are often found mixed with such typically southern species as oaks, chestnut, tulip tree, cucumber tree, yellow buckeye, umbrella and mountain magnolias, Frazer fir, Carolina hemlock, table mountain pine, shortleaf pine, and Jersey scrub pine. As many as fifty forest types have been described for this area alone. Some of the more important are listed as follows:

Swamp *	White pine-hardwood
Spruce flat	Birch-poplar
Spruce slope	Jack pine
Birch-beech-maple-hemlock	Red pine
Balsam fir	White pine
Pitch pine	Scrub oak

The Southern Forest is confined roughly to the Coastal Plain along the Atlantic Ocean and the Gulf of Mexico, extending from southern New Jersey to eastern Texas in a belt varying from 150 to 250 miles inland. The forest consists mostly of pure stands of shortleaf, longleaf, and loblolly pines, sometimes in mixture with other southern pines. These occur mainly upon the sandy and gravelly soils typical of the Coastal Plain. Inland some of these pines are found in mixture with oaks and minor hardwoods. On the heavier soils of the bottomlands, especially along rivers, a large variety of hardwoods grow with cypress. On the border of swamps and along the bottomlands are such valuable hardwoods as oaks, hickories, ashes, gums, sycamore, cottonwood, magnolia, elm, walnut, maples, and others, their distribution depending largely on soil drainage.

* Forest types are generally named for the predominant tree or trees, or for the prevailing topographic feature of the type.

Taking the region as a whole, with its varying soil and topographic conditions, the forest is extremely complex. While the prevailing pine type is simple enough in composition, on the edges of this pinery great diversification occurs. There are not only sands of varying degrees of moisture, but also sandy uplands, sandy ridges, limestone soils, heavy prairie soils, coral reefs, shell hammocks,

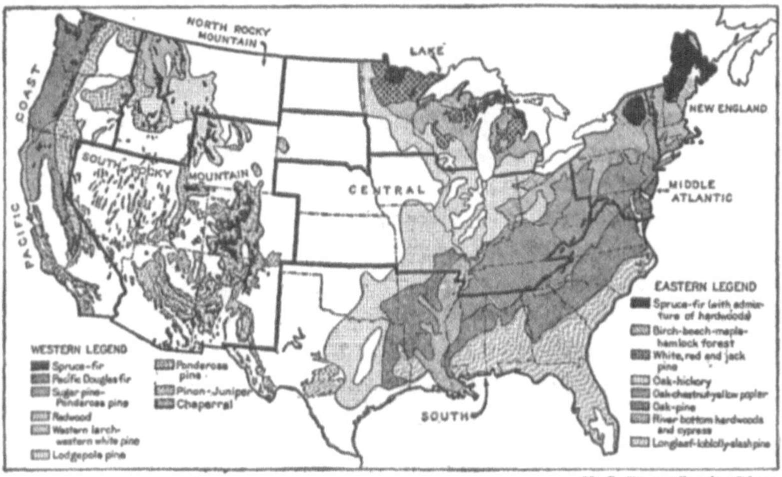

U. S. Forest Service Photo

FOREST REGIONS AND PRINCIPAL TYPES OF FOREST
(Based on forest cover map "Forests of the United States" by Zon and Shantz)

coastal marshes, overflow river bottoms, well-drained river bottoms, and a large variety of swamps. Each of these soils and conditions gives rise to certain vegetative types, many of which have trees in them. One could easily recognize over thirty sub-types. Moreover, the variety of species is also very large. It is said that the forests of southern Florida alone contain over 175 different kinds of woods. The principal forest types from the forest-production standpoint are:

Shortleaf pine	Pine-oak
Longleaf pine	Swamp
Loblolly pine	River bottom
Slash pine	Tropical forest

The Central Hardwood Forest occupies the region between the northern and the southern forests just described. It extends eastward from the prairie fringe forests of western Minnesota, eastern Iowa, eastern Kansas, and central Texas to the base of the Appalachian Mountains and it includes a narrow belt along the eastern base of these mountains extending northeastward to southern New York and Connecticut. The region contains a great variety of species, the state of Illinois alone recording over a hundred native trees.

The latitude of the Ohio River roughly divides this region into a northern and a southern sub-region. North of the Ohio River the western half is predominantly an oak-hickory forest which extends to the fringe forest of the prairie region, while the eastern half is composed mainly of a mixture of beech, birch, and maple. South of the Ohio River the predominant types are oak, chestnut, and yellow poplar (tulip tree), merging into the oak-pine belt of the southern forest and with the numerous species of the northern forest on the lower slopes of the southern Appalachian Mountains. Both of these sub-divisions contain a large number of forest types, usually designated as river bottom, upland, lower slope, upper slope, flat, ridge, cove, and in similar topographic terms. The bulk of the timber is held in the form of small woodlands in contrast to the large holdings which are the rule in the other two regions just described.

There are probably about 150 species of trees that grow in this region which are important to forest production either as principal species or as associates. These can be divided in a general way into four large geographic groups. The first group may be called typical of the entire region, occurring both north and south of the Ohio River: about twelve species of oaks; five species of hickory; chestnut; black walnut; butternut; cottonwood; sycamore; honey and black locusts; buckeye; sweet gum; blue ash; and about fifteen others of lesser importance. Then there is a group of species typical of the northern forest which is also found throughout most of the central hardwood region: sugar, black, red, and silver maples; black birch; beech; basswood; white, slippery, and rock elms; ironwood; white ash; black cherry; and a few others of lesser importance. The third group includes several northern species which overlap into the central hardwood region: yellow, gray, and paper birches; quaking and large-tooth aspen; balsam poplar; black ash; fire cherry; hemlock; white pine; and several other northern evergreens. The last group may be called typical southern species which reach their best development south of the Ohio River: yellow poplar; cucumber tree; umbrella tree; persimmon; pecan and water hickory; cow, black jack, overcup, Spanish, Texas, willow, and water oak; yellow wood; winged elm; osage orange; catalpa; shortleaf, pitch, table mountain, and Jersey scrub pines; bald cypress; and southern white cedar.

Six general forest types of this region may be distinguished, although, as has already been stated, each could be further subdivided into sub-types depending upon local topographic conditions:

Oak-hickory	Oak-pine
Birch-beech-maple	River bottom (North)
Oak-chestnut-yellow poplar	River bottom (South)

The Rocky Mountain Forests for the purposes of this discussion occupy the lower, middle, and upper slopes of these mountains between the elevations of 1,500 and 6,000 feet in the north and from 5,000 to 12,000 feet in the south. Since there is a marked increase in rainfall and relative humidity and a decrease of average yearly temperatures with altitude, the general forest types are found in altitudinal belts one above the other, from the lower slopes bordering the semi-arid plains and valleys to the limit of tree growth at timber line. The commercial forests are entirely coniferous, the species of poplars, birches, willows, oaks, and alders being of no value as lumber.

We shall group the species of this region into those that are found in the northern states of Idaho and Montana; the central states of Wyoming, Colorado, Nevada, and Utah; and the southern states of Arizona and New Mexico. Trees that are characteristically northern are western white pine, white bark pine, western larch, alpine larch, western hemlock, mountain hemlock, western red cedar, western yew, white spruce, and grand fir. One species, lodgepole pine, is both northern and central in its distribution. Two unimportant species are typically central, namely, Utah juniper and bristlecone pine. Several species are both central and southern in range: blue spruce, white fir, one-seed juniper, edible pinyon (*pinus edulis*), and single-leaf pine. Characteristically southern species are the cork fir, alligator juniper, Mexican white pine, Mexican pinyon, Arizona pine, Apache pine, Chihuahua pine, and the Arizona and the smooth cypress. Besides these there are several very important species that range more or less throughout the whole region: Douglas fir, western yellow pine, limber pine, Engelmann spruce, alpine fir, Rocky Mountain red cedar, and the common juniper.

The Rocky Mountain trees mingle with the Pacific forest on the west and with the Atlantic forest on the east. Thus in the north such typical Rocky mountain species as lodgepole pine, Engelmann spruce, alpine fir, western white pine, limber pine, and western larch are mixed, often in the same type, with Pacific Coast species like western hemlock, grand fir, mountain hemlock, western red cedar, and western yew. In the Black Hills of Wyoming and South

Dakota an interesting mixture of the Rocky Mountain flora with the Atlantic flora occurs. Elevations in these hills range from about 3,000 to a little over 7,200 feet. Western yellow pine occupies the lower slopes up to about 4,000 feet and white spruce, an Atlantic species, occupies the upper slopes. In these pine and spruce forests, mainly along the streams, such eastern species as paper birch, burr oak, box elder maple, white elm, red or green ash, ironwood, cottonwood, and red cedar may be found with the Rocky Mountain poplar and the western birch.

Listed in the order of occurrence from the lower to the higher elevations, the principal Rocky Mountain types are:

NORTHERN	CENTRAL	SOUTHERN
Juniper	Pinyon-juniper	Pinyon-juniper
Western yellow pine	Oak	Oak
Lodgepole pine	W. yellow pine	W. yellow pine
Western white pine	Lodgepole pine	Douglas fir
Douglas fir	Douglas fir	Engelmann spruce
Cedar-grand fir	Bristlecone pine	White fir
Cedar-hemlock-grand fir	Engelmann spruce	Aspen
Larch-Douglas fir	Aspen	Sub-alpine
Engelmann spruce	Sub-alpine	
Mountain hemlock		
Sub-alpine		

The Pacific Forest.—The altitudinal distribution of the Pacific forests varies considerably from the northern part of Washington to the southern extremity of California. The forests range increasingly higher from north to south and from west to east.

The principal forests of the Olympic Mountains in Washington occur between sea level and 5,500 feet; in the southern California coast ranges, forests occur between 5,000 and 10,800 feet, although some alpine species in the San Bernardino Mountains climb to 11,800 feet. In the northern Cascades, forests range from sea level to about 7,000 feet, but white-bark pine on Mt. Rainier climbs to about 8,200 feet. In the Blue Mountains of Oregon, forests occur between 1,500 and 8,500 feet; in the Cascades of southern Oregon from 1,300 to 9,000 feet; in the northern Sierra Nevada from 2,500 to 9,500 feet; in the Warner Mountains from 6,000 to 10,000 feet; and in the southern Sierras from 3,000 to 11,000 feet, although limber and white-bark pine sometimes climb to 12,000 feet.

Hardwood forests, principally oaks, occur below these coniferous forests at the western base of the Cascades in Oregon and in the valleys of California. Open, scattered stands of digger pine and juniper are also found in the foothills of the interior valleys of California below the yellow pine belt. In the southern coast ranges and in the foothills of the Sierras a belt of permanent chaparral usually occurs below the digger pine.

The Pacific Coast Forest Region, although the smallest in point of area, has a great variety of climate and topography, since it embraces about sixteen degrees of latitude and elevations ranging from below sea level to 14,500 feet (Mt. Whitney). Over 140 species of trees are found in this region, besides over 100 species of xerophytic shrubs found in the chaparral of California. Most of these trees are of commercial importance or are associated with species that are; the chaparral is of great importance for watershed protection to prevent rapid run-off, erosion, and avalanches.

The Pacific forest contains a large number of species that are found in the Rocky Mountain forest, besides a considerable number that have a distinct Pacific Coast range. Of the thirty-four species of conifers that occur in the Rocky Mountains, twenty-one are found in the Pacific area. These are western white pine, limber pine, white-bark pine, single-leaf pine, bristlecone pine, western yellow pine, lodgepole pine, alpine larch, western larch, western hemlock, mountain hemlock, Douglas fir, western red cedar, western yew, Engelmann spruce, alpine fir, grand fir, white fir, common juniper, Rocky Mountain juniper, and Utah juniper. To these must be added the distinctly Pacific Coast species of which there are twenty-nine. This list includes ten pines, six cypresses, four firs, three spruces, big tree, redwood, incense cedar, western juniper, California juniper, and California nutmeg.

But it must not be assumed that these fifty evergreens are distributed through all the Pacific Coast states. As a matter of fact, only nine have such a wide range, namely, western yellow pine, Douglas fir, white-bark pine, western white pine, lodgepole pine, mountain hemlock, western yew, western juniper, and common juniper. There are eight which are found in Washington and Oregon but not in California: western larch, alpine larch, Engelmann spruce, alpine fir, amabilis fir, noble fir, Alaska cypress, and Rocky Mountain juniper. There are sixteen species which occur in California and in no other state in the Union: four-leaf pine, fox-tail pine, Torrey pine, digger pine, Coulter pine, Monterey pine, pricklecone pine, bigcone spruce, bristlecone fir, bigtree, Monterey cypress, Gowen cypress, dwarf cypress, Macnab cypress, California juniper, and California nutmeg. There are seven California trees which would also be included in the previous list were it not for the fact that they overlap into southern Oregon: sugar pine, Jeffrey pine, knobcone pine, white fir, red and Shasta fir, redwood, and incense cedar. There are also six Washington and Oregon species which overlap into northern California: Sitka spruce, weeping spruce, western hemlock, grand fir, western red cedar, and Lawson cypress. Finally there are four species which do not occur in Washington and Oregon, but which are found in California and in some of the Rocky Mountain states: limber pine, single-leaf pine, bristlecone pine, and Utah juniper.

Besides these fifty evergreens there are also a large number of hardwoods on the Pacific slope. There are sixteen species of oaks, ten species of willows, four of ash, four of maple, two of birch, three of alder, one walnut, and many others scattered through other genera, bringing the total to eighty-eight. One palm and two yuccas bring the grand total of all species for the Pacific states to 141. Of this number, 125 occur in California.

It is true that many of these species are of minor importance. Probably only about one-half of the evergreens are of value for lumber, while few of the hardwoods are so used. However, the minor species in the aggregate may be of great importance in some localities for fuelwood, nuts, tanbark, drugs, chemicals, and other by-products; for watershed and streambank protection; for livestock browse; as food, shelter, and breeding grounds for wild life; as nurse crops following the cutting of valuable species; and for scenic and recreational value. In a dry region like this one, many of them grow where timber trees cannot thrive.

Listed in the order of their general occurrence from the lower to the higher elevations, the following forest types have been recognized:

NORTHERN (Oregon and Washington)	SOUTHERN (California)
Juniper	Redwood (along coast)
Oak	Oak (interior valleys)
Western yellow pine	Juniper
Sugar pine-yellow pine (southern Oregon only)	Pinyon-juniper (eastern Sierras)
	Permanent chaparral (southern)
Lodgepole pine	Digger pine
Douglas fir	Yellow pine
Douglas fir-larch	Jeffrey pine

Western hemlock
Western red cedar
Fir
Mountain hemlock
Sub-alpine

Temporary chaparral (also in Oregon)
Sugar pine-yellow pine
Sugar pine-fir
Lodgepole pine
Douglas fir
Fir (either white or red fir)
Sub-alpine

CHAPTER III

RECREATION FOR THE PEOPLE

BEFORE THE DEVELOPMENT of the automobile, few people were able to visit the forests for pleasure or relaxation. Only the hardy sportsman or naturalist could penetrate the forest trails, pitch camp by a lake, or climb to the top of a mountain. But within the past twenty-five years the situation has changed: with three million miles of improved roads and nearly thirty million cars, travel for recreation has increased amazingly. Americans are bent on seeing their great continent as conveniently and at as little expense as possible. Since this movement is bound to develop at an ever-increasing pace, intelligent planning for the future is urgently needed.

Before the temporary let-down caused by the war, forest recreation had a phenomenal growth. National Parks and National Forests had about 3,500,000 visitors in 1917; in 1937 they had over 40,000,000. In 1910 America knew practically nothing about skiing; shortly before the outbreak of the war, there were estimated to be over 3,000,000 ski enthusiasts. They were spending more than $20,000,000 annually for skis and accessories, clothes, instruction, transportation, lodging, and incidentals. They brought into existence steam-heated hotels in summer resorts, ski-tows, ski jumps, ski schools and snow trains. One large winter-sports establishment in New England represents an investment of over a million dollars.

The total income from all forms of forest recreation amounted to about $2,000,000,000 each year. In the six New England states the annual returns from recreational investment (largely concentrated among the forested lakes and mountains) reached $500,000,000. So large an income was no doubt due to the fact that this region adjoins the most densely populated part of the country.

The total investment in recreational facilities all over the country can be only roughly estimated. Nor is it possible to appraise the taxable wealth which has been added by forest recreational industries in the United States, or the amount of wages paid by these facilities. Needless to say, we are dealing with a financial structure of large proportions. Finally, it must be remembered that the most important results of forest recreation cannot be valued in terms of money. These are concerned with the vitality, health, and happiness of 130,000,000 people.'

Aldo Leopold, Robert Marshall, and L. F. Kneipp were the first to call attention to this huge industry and its future possibilities. It was they who first conceived a coördinated plan for the organization and management of the country's forest recreational wealth. The comprehensive and far-sighted plan which they devised forms the basis of the following discussion.*

Value and Purpose of Recreation

It is hard to express in words all that we see and feel during a vacation out-of-doors, especially as each person finds something different in such an experience. Yet there is much in common between the joy of a mountain-climber as he scales a rugged peak, and the excitement of an old lady as she glimpses a porcupine from the hotel porch. Every person who goes to the woods may find, if he wishes, physical and mental relaxation, sensory stimulation, the chance to study nature at first hand, the adventure of facing the wilderness, and a new appreciation of beauty. From all this will come health, happiness, and inspiration.

The biologist will tell us that recreation "re-creates" mind and body by stimulating nerve and muscle fibers that are not used in our everyday work. We are then given a chance to rebuild and store up energy for the rest of the year. Our mechanized society with its monotonous routine also makes recreation necessary at regular intervals to relieve the nervous tension. The man who sits at a desk needs exercise, fresh air, and sunshine. The factory worker needs rest and freedom from noise. Thus it would seem that to some people recreation means getting away from speed and excitement, and to others it means a limbering-up through greater activity.

Who can describe the effect of recreation on the mind—the sense of ease and mental relaxation? Many people declare that forest, lake, and stream give them a new confidence, will-power, and determination. There is a rediscovery of the real "self," through which we achieve perspective, think more boldly, make plans, and reach decisions. We know exactly what to do about problems that face us for the coming year.

Much of this spiritual and moral rebirth is doubtless due to the

* *Forest Outings,* U. S. Forest Service, 1940.

sensory stimulation we get from our new environment. The cool mountain air, the feel of moss underfoot on forest trails, the sound of bird songs in the morning, the roar of a cataract, the smell of trout sizzling at the campfire—all these open a new world of aesthetic understanding. Our sense of color is sharpened as we learn to distinguish the pale gleam of white-birch bark from the golden patina of yellow birch—the soft silvery green of white-pine boughs from the darker green of hemlock. Who can forget the first exhilaration of reaching a mountain top—with bright lakes and checkerboard farms spread out below, and range after range of blue peaks receding into the hazy distance? Deep canyon walls not only display the most delicate pastel shades in their varied strata, but they also record the story of vanished ages. Such scenes as these encourage contemplation and engender humility. Some of our most beloved naturalists—Thoreau, Muir, and Burroughs—have interpreted these things for us in books that promise to be immortal.

It is perhaps to the naturalist—whether professional or amateur—that the forest yields its greatest reward. So long as the mind is awake, it is never too late to begin the study of nature for ourselves. From mouse to mountain lion, from micro-organism to giant sequoia, from the tiny hidden flower of trailing arbutus to the eighteen-inch petal-spread of the wild magnolia in the North Carolina mountains—in all these we find the utmost variety of form and color and function. And the forest is the natural habitat of a large proportion of plant and animal life. We learn to recognize and identify familiar forms and then we seek for rarer ones. All Europe has hardly half-a-dozen species of oak—while the United States has more than eighty! To know even ten of these as old acquaintances, the bark, the leaves, the acorns, is not only a joy but the highest sort of education—for we learn above all that the great principle of *order* prevails throughout the natural world.

Many naturalists are also collectors, and their specimens often find their way into college and city museums, each item tagged with its name and the place where it was found. Others hunt with the camera, spending tireless patience and ingenuity in photographing the shyest animals and birds, their breeding grounds, nests, and habitats. The biologist hunts with his microscope, prying into the life-cycle of teeming organisms from pond and stream. All are moved by that distinctive passion of humankind—intellectual curiosity.

An even broader appeal, however, is found in the challenge of the dark virgin forest and the uncharted wilderness to man's spirit of adventure. There is in us a deep-seated instinct to lay aside the trap-

Photo New York Conservation Department

EAST: LAKE GEORGE AND ADIRONDACK MOUNTAINS

Forests and landscapes of exceptional beauty, such as this in New York State, must be preserved for posterity.

U. S. Forest Service Phc

WEST: GIANT REDWOODS IN STATE PARK, HUMBOLDT COUNTY, CALIFORN

These are of the Coast Redwood species, tallest trees in the world, though not so massive as the Sierra Redwoods.

U. S. Forest Service Photo; Duluth *Herald* and *News Tribun*

NORTH: SUPERIOR NATIONAL FOREST, MINNESOTA

The country's national forests contain thousands of lakes suitable for fishing.

Photo by Bayard Wootten

SOUTH: MOUNTAIN STREAM NEAR HIGHLANDS, NORTH CAROLINA
There are still a few tracts of virgin timber in the Southern Appalachians, especially in the Great Smoky Mountains National Park.

U. S. Forest Service Photo

HEAVY STAND OF DOUGLAS FIR, WASHINGTON STATE

This shows some of the most magnificent forest in the world. Trees are often 10 to 12 feet in diameter and 250 feet high.

U. S. Forest Service Phot

BEECH AND HEMLOCK, ALLEGHENY NATIONAL FOREST, PENNSYLVANIA
Few virgin stands remain in a state that was once heavily forested.

U. S. Forest Service Photo

VIRGIN STAND OF TULIP TREE AND WHITE OAK, ILLINOIS

Such stands are rare today. The white oak is our most valuable hardwood, and the tulip tree (known to the trade as "yellow poplar") is useful for lumber and ornamental planting.

Photo by Cowling, Courtesy U. S. Forest Service

PINE, SPRUCE, AND FIR, SANTA FE NATIONAL FOREST, NEW MEXICO

U. S. Forest Service Photo

VIRGIN STAND OF RED PINE, CHIPPEWA NATIONAL FOREST, MINNESOTA

Note flourishing young growth under the mature trees. Only a remnant remains of the once magnificent forests of the Lake States.

U. S. Forest Service Photo

PONDEROSA PINE, DESCHUTES NATIONAL FOREST, OREGON

Also known as western yellow pine, this is the most widely distributed tree in the West. Next to Douglas fir, this species holds the largest reserve of standing timber in the United States.

U. S. Forest Service Photo

LODGEPOLE PINE, KANIKEE NATIONAL FOREST, IDAHO

Note the two-storied character of this forest. The under-story is composed mostly of the Douglas fir, a tree highly important to the lumber industry.

U. S. Forest Service Phot

VIRGIN STAND OF REDWOODS NEAR CRESCENT CITY, CALIFORNIA
Many of these trees are over 1,500 years old and range up to 365 feet in height.

Photo by E. S. Shipp

VIRGIN STAND OF BEECH, PISGAH NATIONAL FOREST, NORTH CAROLINA

U. S. Forest Service Pho

HANDSOME STAND OF LOBLOLLY PINE, FRANCIS MARION NATIONAL FOREST, SOUTH CAROLINA

These mature trees were later selected for cutting. Under good management, forests reproduce and perpetuate themselves indefinitely.

U. S. Forest Service Photo

BEARTOOTH LAKE CAMPGROUND, SHOSHONE NATIONAL FOREST, WYOMING

There are more than 3,800 picnic and campground areas in our national forests.

NORTH FORK CAMPGROUND, SAWTOOTH NATIONAL FOREST, IDAHO

There are over 70,000 miles of streams suitable for fishing in the national forests.

U. S. Forest Service Photo

Photo New York Conservation Department

LAKE GEORGE, NORTHERN NEW YORK

Nearly three million acres of forest land has been acquired by New York State for watershed protection and recreation.

BLUE MOUNTAIN LAKE IN THE ADIRONDACKS

About 700,000 people use the recreational facilities of the New York Conservation Department each year.

Photo New York Conservation Department

pings of civilization and go back to the days when man had to rely on his own efforts to keep alive. Like the cave man, we still enjoy matching our wits against the forces of nature. We not only accept but delight in the necessary chores of cooking, washing clothes, and making shelter, under conditions which would not be tolerated at home. We crawl up a steep precipice, cross the mountain torrent on a narrow log, struggle through a maze of fallen timber without compass or guide, and enjoy all these hardships as an exhilarating pleasure. Words cannot express the triumph of landing a ten-pound salmon or of dropping a seven-hundred-pound elk in his tracks.

Fortunately, there are many ways of finding this health and happiness in the woods. Some people take their scenery at sixty miles an hour, and others, with more energy and discrimination, want to live in intimate contact with the forest itself, taking the poison ivy and mosquitoes as part of the game. To satisfy everyone, there is roadside camping and wilderness camping; trips of a few days or of many months; the pack and blanket-roll or the luxurious trailer, to suit the traveler's whim and financial status. One may go on foot or horseback, in canoe, sailboat, motorboat, or airplane—and in winter, the vacationist may travel by snowshoe, ski, toboggan, or iceboat, depending on whether he prefers three miles an hour or up to a hundred.

One thing must not be forgotten: forest recreation should be available at moderate cost. Nearly half of all American families have an income of less than a thousand dollars a year.* This fact should not deprive them of their right to enjoy our natural heritage. To supply such an opportunity for millions of Americans is a huge assignment, which should be shared by local, state, and federal government, with no thought of the profit motive.

Types of Recreational Areas

Any rational plan for forest recreation must recognize several distinct types of areas which have individual characteristics as to size, kind of scenery, kind of use, and administration. They are so different in character and function that separate recreational programs must be developed for each.

Superlative areas are regions of exceptional scenic beauty which are so unusual that they are of national importance. Examples are such areas as the geysers of the Yellowstone, the Grand Canyon, the Big Trees, the Carlsbad Caverns, the Mammoth Caves of Kentucky,

* Based on consumers' incomes for 1935-36, cited in *Forest Outings* (U. S. Forest Service, 1940), p. 293.

the Great Smoky Mountains, and many others just as familiar. Most of these "scenic wonders" have already been set aside as National Parks. Some are located in National Forests, national monuments, or state parks. It is not necessary to justify their retention as recreational areas—it is only necessary to safeguard them from commercial exploitation and see that they remain accessible to all the people. A few of our superlative areas remain in private ownership, and these should be acquired by public agencies.

Primeval areas might be called "wild nature areas." They include tracts of virgin forests unspoiled by human hands. For the most part they are in much the same condition as when the white man first settled America. Such areas are of great scientific value in planning methods of cutting and reproduction for the perpetuation of similar forest types. They are helpful in understanding the many laws of forest growth, development, succession, and distribution. Of no less importance is the great beauty of these remnants of our original forest. In southern Michigan, for instance, are oak groves in which many of the trees range from three to five feet in diameter. One cannot travel in the Adirondacks without wondering what the original virgin white pine stands looked like in all their unspoiled magnificence. Here and there one can still find isolated specimens over five feet in diameter and from 150 to 175 feet in height, mere suggestions of the original growth.

Both the Forest Service and the Park Service are making provisions for reserving primeval areas. In the twenty major timber types of the United States over fifty areas have been set aside by the Park Service. The Forest Service and several states have set aside a number of smaller areas for the same purpose. In general, they average in size from 1,000 to 5,000 acres, the size necessarily depending on how much virgin forest is available. The areas may be put to both scientific and recreational use, though the latter will in some cases be greatly restricted. It is of course highly desirable that rapid progress be made in establishing primeval areas in the various forest types; no doubt the possibility of establishing such areas in certain types has already been lost, because of their complete destruction.

Wilderness areas are large regions which have practically no permanent inhabitants or any of the modern means of transportation. Since they preserve the essentials of a wilderness, even though logged off, visitors must rely entirely on their own efforts to travel therein. Roads, settlements, and power lines are barred, but trails and temporary shelters are permissible. The Forest Service and the Park Service have both adopted a policy of reserving such areas from residential use and from penetration by roads. Most of the land involved, on account of its inaccessibility, altitude, and slowness of forest growth would possess little value for the production of timber. Such forests are essentially for the protection of land and water resources.

Marshall adopted a minimum of 200,000 acres as the size for such wilderness areas. He listed thirty-eight established, partly established, or potential wilderness areas, nine of which have 1,000,000 acres or more, nine others 500,000 acres or more, and twenty 200,000 acres or more. The total number of such areas available at the present time is given as sixty-five, totaling some 27,000,000 acres.

The author has camped in some of these areas for months at a time and can say from experience that wilderness journeys, especially if they involve pack animals, packers, cooks, guides, and necessary equipment, are expensive. A month's trip of from three to four hundred miles would entail an expense of from ten to fifteen dollars a day. A very hardy woodsman might take a two weeks' journey at comparatively small expense. Wilderness areas already established in California and Wyoming with which the author is familiar, often require a journey of from 100 to 150 miles from a town to reach the area and then an additional 25 to 50 miles by trail to get into the wilderness part of it.

The three types of recreational areas just described are not the kind that will receive the most frequent use, chiefly because of their inaccessibility and the expense involved. To meet the needs of the average automobile tourist *roadside areas* are being rapidly established. Most of the visitors do not penetrate far into the forest but enjoy the scenery from their cars. They are naturally concerned with the beauty of that part of the landscape which they can see from the road. For these people it is important to preserve from scenic damage timbered strips along the more important highways. It is thought that this plan would involve setting aside about 100,000 miles of roadside strips. On these strips existent scenery must be protected, some will have to be restored by planting and landscaping, and others must be improved by the elimination of billboards and unsightly houses. This class of recreational area will include also strips of forest left along lakes, rivers, and other scenic water routes.

Camp-site areas will provide not only comfortable and convenient land for camping, but also will curtail the damage which results from indiscriminate camping in the woods. By providing fireplaces, tent sites, drinkable water, and firewood on restricted areas suitable for such activities, the recreational values of the entire forest are safeguarded. Perhaps the greatest value of regulated camping of this kind is the great reduction of the fire hazard. In the more populous eastern states thousands of these camp-site areas are already in existence, and over certain week-end holidays their capacities are taxed to the utmost, attesting to their popularity. There is often a resident caretaker for the summer months, and campers are required to register and give their auto license numbers.

Residence areas provide forest land for private homes, hotels, resorts, over-night cabins, and sanitoria, and stores and services of one kind or another. In many cases these are private areas rented for the summer months. Many publicly owned areas also exist. The Forest Service in 1931 furnished summer home sites for about 500,000 special-use permittees and their guests. Many of these were near large western cities.

Lastly, *outing areas* are designed for those travellers who prefer more intimate contact with the woods. Sooner or later, enjoying scenery at forty miles an hour from a public highway gives way to exploring the woods and streams beyond, where one can camp without dust, noise of motors, and other modern annoyances. These outing areas will often have special value as buffers for superlative and primeval areas. Such areas will have accessibility and will be a compromise between the wilderness areas and the camp-site areas. The necessity of having these areas owned by the public, so that timber use and recreational use can be combined, is obvious from the fact that most private owners do not as yet practice sustained-yield forestry.

While the above-mentioned recreational areas are rather distinct in function and in management, it happens quite often that certain forest areas offer a combination of some of these facilities. A state like New Hampshire is a good illustration of this. It has over 100 public recreational areas of many types and descriptions, besides a large National Forest which provides many public forest camps. It has such superlative areas as the "Old Man of the Mountains," the Flume, and Lost River. It has primeval areas such as the Bowl, which contains a heavy stand of virgin spruce and hardwood. Wild River forest camp might approach what is termed a wilderness area. There are many state beaches without camping, state camp grounds without bathing, and state camp grounds with bathing, besides forest fire lookouts, scenic drives, experimental forests, game management areas, and state fish hatcheries and rearing pools.

The Wellington State Forest and Recreational Area, up in the lake region of New Hampshire, consists of about 100 acres of dense Norway pine from 14 to 18 inches in diameter with an under-story of hemlock, aspen, maple, blueberry, beach plums, and other plants, bordering a magnificent mountain lake. The beach, one-half mile long and quite broad, is of clean white sand, forming a crescent flanked on the left by Sugar Loaf Mountain and on the right by a long, pine-covered peninsula. In the distance are the White Mountains. The whole area was donated to the state by a public-spirited citizen. The State Forest serves also as a bird sanctuary.

This area has all the facilities of a public playground. There are fireplaces, piped water, and refreshment stands. There is a ranger cabin, three lifeguards, a diving raft, and a wharf for speedboats and hydroplanes. There is also a well-marked safety zone near shore to protect the children. In the woods back of the beach are bathing houses, rest rooms, and pavilions where one may sit down for lunch or to enjoy the view. The only charge is 25 cents for parking. The area is closed at night, and no camping is permitted.

National Forest Vacations

"Where away? Take the nearest road to the National Forests for the perfect vacation. They are open to everyone. They are yours to enjoy." Such is the enticing invitation extended in a recent Forest Service bulletin entitled "National Forest Vacations."

We have 160 National Forests, comprising 176,000,000 acres and distributed over 40 states. For variety, extent, and interest they are unequaled anywhere else in the world. From any part of the country, some National Forest is within a day's drive by automobile, and the opportunities for recreation are therefore both informal and inexpensive.

Entrance to a National Forest is marked by a sign with the figure of a pine tree centered on a shield. Those who are unfamiliar with the region may look up the nearest supervisor, ranger, or guard, who will be glad to give information or show the way. What one can see and enjoy is limited only by the amount of time at his disposal. There are 137,000 miles of forest highways and over 150,000 miles of hiking and riding trails, on which hardly a mile is without some scenic view, to say nothing of some 70,000 miles of streams and thousands of forest-framed lakes. Practically all National Forests offer opportunity for picnicking, camping, hiking, fishing, hunting, and motoring; many of them have lakes and streams for swimming, boating, and canoeing. Winter sports are enjoyed on the miles of picturesque snow trails. As pointed out in the bulletin mentioned above, the scenery, climate, and topography are as diverse as any in the world, as the following quotation shows:

"Ranging from the mysterious cypress swamps and clumps of sand pines of the South to the alpine meadows and great fir and spruce forests of the Rockies and the Pacific Northwest and from the spreading hardwoods of the Central States and the Appalachian Mountain region to the spectacular, towering redwoods of California, the national forests include timber types representative of every forested region of the country.... There are dense, humid forests and parklike semiarid forests. With the exception of a few units in the deep South and the Lake States, the national forest system is synonymous with hilly and mountainous terrain—characterized by cool or cold nights and warm, sunny days in summer; salubrious, colorful fall seasons; and cold, snowbound or blustery winters."

Picnic and campground facilities provided are sufficient to care for 240,000 people at one time. Some very popular areas have as many as 100,000 visitors in one year; others, more remote, hardly 200 a year. There are over 3,800 camp sites that have been specially selected and improved for the convenience and enjoyment of picnickers and campers. The National Forests have over 13,000 summer home sites leased at the rate of from $15 to $25 per year. Summer homes are erected by the lessees. There are also over 800 resorts, mostly privately owned and operated.

Those who yearn for solitude in the wilderness can choose from about seventy such areas set aside in vast undeveloped regions. These areas are visited yearly by organized pack trains or by single saddle and pack-horse outfits. Inexperienced wilderness travellers are warned to use the services of a guide. Since 1933 the American Forestry Association has offered an opportunity for many to vacation briefly in the unspoiled wilderness through their expeditions known as "Trail Riders of the Wilderness." These are carefully organized with experienced guides, packers, wranglers, and cooks. A physician also accompanies each party. Nearly 600 men and women have already participated in over 40 expeditions. Some typical trips have been as follows:

WILDERNESS AREA	STATE	DURATION	COST	POINT OF DEPARTURE
Sawtooth	Idaho	10 days	$124	Sun Valley
San Juan	Colorado	9 "	138	Durango
Gila	New Mexico	10 "	115	Silver City
Snowmass	Colorado	10 "	135	Glenwood Springs
Sequoia	California	10 "	140	Mineral King

But with all this available enjoyment, the vacationist must be reminded that the National Forests are not exclusively for vacation purposes. Under the multi-use policy of the federal government, the forests are strictly business institutions designed to yield the greatest benefits from all uses to the greatest number of people. In other words, they not only provide vacations for more than thirty millions annually, but also a livelihood for more than a million persons. Several billion feet of lumber are cut on them annually for America's industries; they furnish range for ten million head of live stock; and they are the source of domestic water for six million city people. All of which means that they are the peoples' forests and each individual has a share in their ownership and a responsibility to use them wisely.

We repeat: these forests are for use without charge. Your only ticket for admission is the solemn pledge:

1. I will not be careless with fire in the woods; I will drench my camp fire before leaving it.

2. I will leave the forest camp in as neat and clean a condition as I would like to find it.

3. I will not mark, hack, or deface what Nature gave me to enjoy.

4. I will not stray off the well-posted forest roads without maps, compass, or reliable guide.

Future Administration

The uneven distribution of the population of the United States is a fact of great significance in considering the location of needful facilities for outdoor recreation.* The western states with their large area and abundant recreational facilities have only a small population, while the states of the northeastern quarter of the country, with only 13 per cent of the total land area and comparatively few outdoor recreational facilities, have nearly 57 per cent of the population. The greatest inequality in this respect is found around the large urban centers. The 47,000,000 inhabitants of cities of 30,000 or more are crowded on only 0.002 per cent of the total land area of the United States.

Using the ratio of one acre of recreational area for each hundred persons of the city population—suggested by experts as a minimum standard—the cities having 5,000 or more population have less than half the desirable minimum of recreation space. Hence, at present there is a heavy burden of recreational use placed upon facilities in and near the urban centers. City parks and playgrounds, park-

* National Resources Board *Report,* 1934, pp. 144-47.

ways, near-by metropolitan areas, county and state parks, and community and state forests are woefully inadequate to meet the needs of city dwellers. Clearly this shows the need for more facilities of this kind.

Looking at the problem as a whole, there are six separate administrative agencies that will manage recreational areas, and each one seems to be well adapted to handle its particular type. In the *National Parks* we shall find most of the superlative areas. In these parks commercial exploitation is barred and road-building is reduced to a minimum. As a result these areas are apt to be kept in their primitive condition. The *National Monument system* of the Department of Interior administers botanic reservations (Muir Woods) and wild animal reservations, as well as areas of historic and geologic interest. They are administered in a manner similar to that of the parks. *The National Forests* as now administered are well adapted to the management of all types of recreational areas. *State Parks* and *State Forests* should be devoted especially to the recreational and other interests of the states in which they lie. For the most part they will be small and scattered, and will be in the nature of roadside, camp site, residential, and outing areas. *County* and *Municipal Parks* are adapted for outings of local citizens and will be in great demand. They should include camp site, residence, and outing areas. Lastly, *private holdings* will be adapted to the types of recreation which will yield a profit. These are generally residence and camp site areas.

Rough approximations made by Marshall as to the total area of forest land that will ultimately be needed by all types of recreational use are as follows:

KIND OF AREAS	ACRES
Superlative areas	3,000,000
Primeval areas	9,500,000
Wilderness areas	10,000,000
Roadside areas	4,000,000
Camp site areas	1,500,000
Residence areas	6,000,000
Outing areas	11,000,000
TOTAL	45,000,000

Already 11,000,000 acres have been withdrawn from timber use in national, state, and local parks. At least 13,000,000 acres of the remaining 34,000,000 consists of inaccessible forest land of low productivity which for a number of reasons should be set aside for recreational use.

Such large withdrawals of lands for recreational use presuppose a great future stimulation in the use of forests for these purposes. This greater use is based upon five well-defined trends in our social and economic life:

1. Population increase to a point of stability estimated at about 165,000,000.

2. Shorter working hours, with more time for leisure and recreation.

3. Probable rise in the American standard of living, with additional millions of workers seeking outdoor recreation.

4. Increased ease of transportation and the advent of super-highways.

5. Increased mechanization of our industrial life, with increased demand for means of easing the nervous strain.

Finally, in the administration of the smaller and more accessible recreational units like roadside, camp site, and outing areas, it becomes increasingly evident that some effective curb must be put on the abuse of these areas by the public. The cutting of green trees, the barking of mature ones, and the forestalling of new growth by tramping and packing down the forest floor mean the serious impairment and ultimate destruction of these facilities. More careful planning of camp sites, more stringent supervision, a prevention of overcrowding, and a general campaign to educate the public to a proper use and appreciation of the facilities provided are measures which seem essential to the preservation of forest recreation for the people.

CHAPTER IV

RELATION OF FORESTS TO WILD LIFE

HUNTING AND FISHING were deeply rooted in the life of early man. They certainly go back to the Old Stone Age, perhaps a million years ago, when the cave-dweller's first concern was to get enough to eat. He learned to match his physical strength and his ingenuity against those of the wild beast. With experience as his only teacher, he made his crude stone weapons and became a hunter and fisher in order to survive. It is not surprising that so ancient a pursuit has its strong appeal today.

Even during Greek and Roman times the forest was thought of mainly in terms of what game it would yield. Mythology is rich in allusions to the chase—and the forest as a source of meat supply was a conception that lasted well through the Middle Ages. The privilege of hunting was restricted mainly to the nobility, and large tracts of woodland were held exclusively for their use. The foresters of those days were merely gamekeepers, and the welfare of the forest itself was nobody's concern. Not until a substantial part of the European forest had been destroyed did the idea grow up that the forest was an important source of raw material for a rapidly developing industrial civilization and that it was also intimately related to man's water supply.

The early settlers of America depended largely on game and fur-bearing animals for their food, clothing, and cash income. Pelts were valuable, and the profit of fur-trading was one of the chief incentives that led the white man to push into the great western wilderness. This traffic contributed in no small way to the westward march of American civilization. Fur-traders established their posts along the rivers and lakes, and their posts grew into villages linked together by trails that grew into roads.

As agriculture and industry opened up the West, wild life became more and more restricted. At the same time, wild birds and game were still one of the main sources of meat supply. Forest devastation, which received great impetus in the latter part of the nineteenth century, made still greater inroads upon the natural habitats of wild life. It is very likely that the destruction of the forests was the greatest single factor in the decimation of our wild life. Even the passenger pigeon, whose enormous flocks once clouded the skies, is said to have suffered more by the destruction of its breeding and nesting grounds than from pothunters. The grazing of domestic animals in the West reduced the range of the elk, the buffalo, and the antelope. It is reported that about three-fourths of our midwestern breeding grounds of wild waterfowl have been destroyed, first by the inexorable conquest of the plow, and secondly by the lowering of the water table as a result of forest destruction. Three species of wild duck are practically extinct.

It is not surprising, then, that there is general acceptance of the fact that forest conservation is the first step toward wild-life conservation. We must use plow lands to raise crops and we must use grass lands to feed meat animals. The rehabilitation of our forest lands is the only alternative left.

Most European nations—England, France, Germany, Italy, and Switzerland in particular—either own or effectively control forest and water resources. When private enterprise fails, government control begins. There is rigid supervision of stream pollution, of drainage and irrigation works, and of forest cutting on mountain slopes. Hydro-electric plants are either owned or effectively controlled. Forest wild life is similarly protected and regulated.

In Europe the regulation and control of natural resources is greatly aided by the fact that so large a part of the forest area is government owned. In this country about 70 per cent of our forest area is privately controlled. Of course the separate states control wild life, even though the land may be owned by private individuals or the federal government. The states make and enforce the fish and game laws, but unless the public can control the forest cover for wild life, there is not much hope for our fish and game. The only fishing and hunting areas whose permanence is assured are the publicly owned lands like the National Forests and the various types of state forest reserves.

Wild Life an Annual Crop

Wild life must be handled essentially as an annual crop. The annual "take" should equal approximately the annual increase in numbers. In most cases wild-life management means an increase in wild-life population; in a few cases, where there is an over-stocking of certain species, it means a reduction. Primarily, since food is the basis of most wild-life relationships, it means research in ascertaining the carrying capacity of different areas for different species of wild animals.

The most difficult problem in wild-life management is the control of population. Other problems must concern themselves with the increased use of public lands for wild-life use, the establishment of refuges for wild life exclusively, the general increase of cover and protection in order to increase carrying capacity, and the transplanting of certain species from over-populated areas to new ranges suited to that particular animal. Dr. W. A. Shoenfeld has defined wild-life management as the "production of sustained annual crops of wild creatures in harmony with the major economic uses of the land." It is therefore essentially a problem of land utilization.

The direct and indirect benefits of wild life may be listed as follows:

1. Income derived from hunting and fishing.
2. Income derived from furs, hides, and other by-products.
3. Use of game for food.
4. The checking of harmful animals which would damage crops and property.
5. Utilization in the forest of forage which otherwise would be wasted.
6. Knowledge derived from biological study.
7. Pleasure derived by naturalist, photographer, hunter, and camper.

Data prepared by the Fish and Wildlife Service shows the annual money value of wild life, including the worth of the meat, the destruction of insects by birds, the value of fish caught, hunters' fees, and the amounts spent by hunters and tourists, to be over a billion dollars. The Special Senate Committee on the Conservation of Wildlife Resources reported that over 13,000,000 licenses for hunting and fishing were issued during 1929. The income to states from hunting licenses alone in 1930-31 reached $10,000,000. Moreover, there was a 400 per cent increase in hunting and fishing in the decade from 1920 to 1930. It must also be noted that the sale of hunting and fishing equipment in 1929 amounted to over $25,000,000. The value of the fur industry, which derives the major part of its raw materials from fur-bearing animals of the forest, exceeds $300,000,000 a year.*

The bird life of the forest is the most difficult item to evaluate. Birds

* *A National Plan for American Forestry* (Sen. Doc. 12, 73rd Cong., 1st Sess., Washington, D. C., 1933), pp. 489-525.

destroy insects and animals harmful to farm crops and farm property, and these birds live mainly in the forest. Total yearly damage done by all insects is said to be around $3,000,000,000. The United States Department of Agriculture estimates that insect-eating birds alone are worth around $400,000,000 a year to the farmer and that another billion dollars a year could be saved if bird life were restored to its proper level. Birds alone can help man to win his fight against destructive insects, which are an ever-growing menace to man's food supply.

THE BALANCE OF LIFE

The whole problem of wild-life conservation revolves around the simple lesson of the "food-chain." Every animal feeds either on other animals or on plants. Meat-eaters feed on plant-eaters. At the beginning of the food-chain is always the green plant—the only organism capable of an independent existence, since it is the only one that makes its own food from inorganic materials. The animal at the end of the chain, by its excreta or by the decay of its body after death, puts back into the environment the simple elements and compounds which are again used by the next generation of plants and animals. A plant louse sucks the juice from a tender shoot of a forest tree. A ladybird beetle eats the louse. A robin eats the beetle, and the hawk eats the robin. Excreta of these animals enrich the soil with compounds which help the tree to grow.

Now it is very evident that any factor which interferes with or breaks this chain may lead to serious results. Thus, to use another example, if trout streams are polluted with wastes, the mature trout may not suffer material reduction in numbers at the start, but the wastes may kill the tender green algae attached to rocks in the stream, which supply food for the tiny trout shortly after they hatch out. It is well known that the weakest link in the chain of life of aquatic animals is the very early fingerling stage. Thus our "food-chain" concept is enlarged to include not only the animal and plant environment but also the chemical, the physical, and the industrial environment of living things.

Hence, the biological basis of a sound conservation program is the broad concept known as the "balance of nature." Few of us realize that if any species were left to reproduce uncontrolled, it would soon cover the earth or fill the oceans (if the food supply held out). This overproduction is prevented by other organisms which feed on it and keep it in check. Unfavorable environmental conditions such as floods, storms, droughts, and excessive cold or heat also have their effect. All these factors bring about a balance in numbers so that all animals and plants can live together in one of various habitats without any one of them gaining the upper hand. In other words, animal population under nature's plan is geared to the available food supply.

Nowhere in nature is the struggle for existence and the survival of the fittest better exemplified than under wild forest conditions. Wolves and coyotes feed on deer, the fastest of them catching their prey, the slower ones going hungry. The swiftest deer escape; the weak and sickly ones are caught. This is nature's way of securing a race that is stronger and better adapted to the environment than if the weak and sickly were allowed to reproduce their kind. This numerical balance among animals was being harmoniously maintained when the white man first came to America, as indicated by the accounts of early explorers. Anyone who has read the diaries of the naturalist Audubon can appreciate the diversity and abundance of wild life in the Middle West, even as late as the first quarter of the nineteenth century.

Unbalancing Nature

Since Audubon's time, we have upset the balance of nature in three major ways: (1) By killing off species for food and other economic uses; (2) by modifying or actually destroying the environment of these species; and (3) by introducing new species to America which came to a new environment in which natural enemies were absent. A sound conservation program must seek to re-establish the balance that has been lost, encouraging the desirable species and keeping the harmful ones in check.

Our ruthless slaughter of wild life has been notorious.* The great herds of bison that once roamed the plains have been reduced to a mere vestige preserved from extinction in game refuges. The great auk, the passenger pigeon, and the Eskimo curlew are already extinct through man's rapacity. The grizzly bear and the sea otter have been almost exterminated by hunters. The whistling swan, egret, snowy heron, and roseate spoonbill were brought almost to the point of extinction before being protected and partly restored. The wild turkey and the heath hen have been exterminated from New England, and a large number of species, such as the wolverine, woodcock, condor, American eagle, and several kinds of duck, are approaching extinction.

By destroying vast forest areas, by draining swamps, by flooding

* I. N. Gabrielson, *Wildlife Conservation* (New York, 1942), pp. 184-93.

land, by dumping sewage and factory wastes into rivers and lakes, and in many other ways, man has destroyed the natural environment of plants and animals. Forests no longer give adequate protection and food to game, or sufficiently shade the deep, cold pools where trout can thrive; and plant foods, once abundant in our primeval forests for our herbivorous mammals, are now scarce or not available.

Finally, we have introduced many foreign pests, mostly insects such as the Hessian fly, the gypsy moth, European corn borer, Mexican bean beetle, Japanese beetle, cotton boll weevil, and many others which thrive on our plant life practically unchecked, causing damage that runs into hundreds of millions of dollars each year.

We have had to learn by trial and error, and some of our wild-life laws have been very stupid. Pennsylvania, some fifty years ago, put a bounty on hawks and owls because farmers complained that these birds destroyed their chickens. Nearly $100,000 was paid in bounties within a year-and-a-half—a sum much greater than the worth of the chickens that might have been destroyed. But that was not all. The hawks and owls had fed upon field mice and other vermin which, after the law went into effect, were left free to damage grain crops and other property to the amount of about $4,000,000. Needless to say, the bounty was removed.

The cry was raised in the western states that the coyote must be exterminated because it killed sheep. With gun and poison the animal was slaughtered—but the jack rabbits upon which the coyote also feeds promptly increased in such numbers that the effect on grain, fruit trees, and forest trees was disastrous. Coyotes are the natural enemies of the jack rabbit and keep them in check.

Many other examples of this kind could be cited. They all show that the balance of plant and animal life is a delicately adjusted mechanism which cannot be disturbed without starting a whole series of biological maladjustments. Closely related to this biological balance is the chemical balance of soil and water. Fresh water harbors thousands of species of algae (pond scums), which are the food of a wide variety of animal forms from one-celled micro-organisms to simple vertebrates. Sometimes this life thrives only in water of a definite chemical composition. These organisms, also, are part of an intricate food-chain in which the inter-relations are often very complex. The caustic ashes of a forest fire can easily upset this food-chain in a lake or brook and thereby endanger the food supply of most of the animals mentioned above.

Finally, we are all familiar with the Dust Bowl disaster which resulted from disturbing the physical balance between plant life and soil. In general, plant life and climate are in close physical relationship to give maximum protection to the soil. Once this physical balance is disturbed by removing the protective soil cover, disastrous results may follow. When the protecting sod was removed from rich prairie land in Kansas, Oklahoma, and Texas, to grow wheat during the First World War, the fine soil was carried by the wind by the millions of tons. That was a serious mistake that should not be repeated.

General Classification of Wild Life

There are many different kinds of wild life; each group has its own peculiar life cycle, food requirements, and habitat conditions. Also each group has its own definite characteristics which affect its economic relationships, management, protection, and restoration. Each also has its own research problems. No legislation can be very effective unless it is based on sound scientific knowledge.

Many classifications * of wild life have been suggested, each depending on the point of view of the classifier. Only a general grouping, adapted to the needs of forest conservation, will be attempted here. The present discussion emphasizes the rôle of the forest in the lives of these groups, especially, the resident game animals, the migratory game birds, the insectivorous birds of forest and woodland, the fur-bearers, the predators, and the fresh water fish.

The following distinct groups might be recognized:

1. *Resident Big Game:* deer, elk, moose, mountain goat, mountain sheep, caribou, antelope, etc.

2. *Resident Small Game:* quail, turkey, grouse, rabbits, squirrels, etc.

3. *Migratory Birds:* song birds, insect eaters, non-game water birds, and shore birds.

4. *Migratory Game Birds:* ducks, geese, coots, rails, etc.; important enough to separate from the last-mentioned group.

5. *Fur-bearing Animals:* mostly small mammals, including foxes, beaver, marten, otter, muskrat, skunk, raccoon, opossum, etc.

6. *Harmful and Predatory Animals:* wolves, coyotes, mountain lions, bears, prairie dogs, rodents, etc.

7. *Fish:* principally fresh-water game fish such as trout, salmon, bass, pike, etc.

* Gabrielson, *Wildlife Conservation.*

THE FOREST A HOME OF WILD LIFE

The forest, when properly protected and managed, affords a home *par excellence* for wild creatures. Wild life requires water, food, shelter, breeding places, and protection, and the forest directly or indirectly provides all of these. Over our vast forest areas all kinds of habitats may be found, from the peat bogs of Minnesota to the Everglades of Florida, from the dry oak ridges of the California foothills to the cold, trout-filled lakes of Maine, and from the snow-capped peaks of the Rockies to the semi-tropical cypress swamps of Louisiana. Within each of these habitats differences of elevation, slope, exposure, soil, moisture, and light intensity produce a great variety of food and shelter.

Typical forest big-game animals are the elk, deer, and moose, also such predators as grizzly bear, black bear, mountain lion, and bobcat. Among the fur-bearers that live in the forest are the fisher, marten, wolverine, and beaver. Among the forest-loving game birds are the turkey, blue grouse, and ruffed grouse. Many smaller birds and mammals also live in the forest. Finally, our great forest areas contain most of the trout streams and many of the best fishing lakes in the country.

Forest habitats vary widely in their value as game ranges. Dense virgin forests of spruce, fir, and redwood on the Pacific Coast; the dense spruce and fir forests and the lodgepole pine forests of the Rocky Mountains; and the dense hardwood forests of our eastern states may offer temporary shelter but very little feed for game animals. In general, the crowns of the trees in such old stands are too high to be used as browse. Also dense shade prevents the growth of shrubs and herbs that provide feed. For that reason, open pine forests, forests with occasional openings, and any forest stands with an abundance of ground vegetation both for food and shelter are more valuable for both birds and mammals than old growth. Thus, cut-over and burned-over lands grown up to brush and perennial vegetation are far better game range than these lands planted up to dense, young pine.

However, comparatively few stands of timber the country over are too dense to exclude all wild life. The use of certain methods of management in reproducing forests will aid in producing favorable conditions for game. The selection and the shelterwood methods of cutting are most favorable for aiding the continuous production of browse in the openings produced. Clear-cutting followed by planting or natural reproduction will provide much browse during the early part of the life of the forest, say for twenty to thirty years, but as the climax of forest growth is reached at maturity, wild life will decline and remain at a minimum until the forest is again harvested. It is conceivable that some species of animals might have to migrate while this change was going on. For this reason, the best forest management for wild life would avoid large cut-over areas. It happens that the best forestry practices also aim to avoid such areas.

Timber and wild life can be grown on the same areas by a judicious use of silvicultural methods. Often far greater returns per acre can be produced with both crops than with only one. On the best forest soils, the forestry interests should predominate; on inferior soils and in areas far from wood markets, wild life may predominate. Perhaps, eventually, wild life will be largely relegated to certain forest areas especially favorable to it, as is done in many parts of Europe. Hunting would then be restricted to certain designated areas and would not be allowed to go on promiscuously as at present.

The favorite haunts of small and big game animals seem to be the thickets found along streams, lakes, and swamps, and even at the higher elevations where conditions are sometimes unfavorable for tree growth and the forest is low and scrubby. Woody thickets on the common border of meadows and forests are valuable, as well as the brush and tangled growth bordering fences on farms. Such places are also the nesting places of song and insect-eating birds. Small areas of young growth, scattered through mixed, all-aged forests afford food, shelter from wind, protection from enemies, and breeding places.

Forests also provide several levels or strata suitable for wild life. Some animals live below ground, others on the ground, some in the brush and thickets, and some in the trees. Those living on the ground do not necessarily compete for food. The deer browses higher than the rabbit, and the elk higher than either. The sort of food preferred by different kinds of animals may also make them noncompetitive on the same area. Rabbits prefer low, tender, herbaceous vegetation; deer browse on bushes and shrubs; elk are primarily grazing animals and prefer grass; the moose eats aquatic plant life; the antelope prefers willow; and the mountain sheep live mostly on moss.*

Furthermore, it is important to note that the forest, through its intimate relation to erosion and stream flow, safeguards all water habitats and therefore all wild life in any way connected with them.

* *Forest Conservation: A Guide to the Social Studies* (U. S. Forest Service, 1937), pp. 78-100.

We have found out that the removal of forests over large areas in the central states has lowered the water table, thereby drying up ponds and swamps, with disastrous effects upon fish, fur-bearing aquatic animals, and aquatic bird life. Mountain forests insure the life of streams which harbor our valuable game fish. All animals of the forest drink from these same streams. Cool, shady mountain streams, so essential for trout, can only be provided by a dense forest cover.

Some Wild-Life "Pictures"

Among the most lasting impressions one gets in the outdoor world are those of wild animals seen or heard. Their beauty and grace, their size and strength, arouse our admiration. Most forest visitors do not hunt with a gun. They hunt with the camera if they happen to have one handy at the moment, and if not, they simply stop, look, and listen to admire and study the animal in its native haunts. Who of us has not seen a bird feeding its young in the nest, or a mother robin chasing an owl away from its young, or a raccoon washing its food in a mountain stream, or perhaps a grotesque, prehistoric-looking moose feeding upon water plants?

The various species of deer are probably the most widely distributed large game animals we have. Essentially browsers, they are naturally inhabitants of the forest. Through the year they subsist on the products of the forest. In the fall they feed on acorns or beech nuts, and in the winter they search for fruits left on the trees and shrubs, on the smaller twigs with their buds, evergreen broadleaf vegetation, the inner bark of trees, and green water plants. In the spring and summer they live largely on browse consisting of tender leaves and shoots, and berries and larger fruits. They also graze on clover and other herbaceous growth. The tender sprouts of juniper seem to be their favorite fare throughout the year.

Shy, graceful and lithe, they also have sense organs quickened to detect danger. Keen eyes, a remarkable sense of smell, and swivel ears are characters which have enabled them to survive. Some National Forests have too many deer; others not enough. Transplanting is carried out whenever feasible. Overpopulation is disastrous not only to the young trees of the forest but also, eventually, to the deer themselves. A careful count of the herds must be made from time to time, and the game laws tightened or liberalized in proportion to the food available.

The elk is like the deer but considerably larger, often weighing 600 to 900 pounds. They are bulky animals, and when frightened in the dense lodgepole pine forests of Wyoming, can be heard crashing noisily through the down timber. Their enormous antlers are highly prized and are often the only part of the animal that can be conveniently carried away from remote regions where the game is shot. The National Forests have about 125,000 elk, mostly in the states of Wyoming, Idaho, Montana, and Colorado, though some have been transplanted to Arizona, California, and other states. It is a beautiful sight to come upon a small herd grazing in the grassy parks of northern Wyoming, or standing serenely in the edge of the water. They are good swimmers. In the fall they feed on grass in the lowlands; in the spring and summer they are back again on the higher ranges.

The moose is the mammoth of the deer family. He often weighs from 1,200 to 1,400 pounds, and when standing in the shallow waters of a mountain lake quietly munching large mouthfuls of water plants, is a sight never to be forgotten. The slightest noise will bring him to attention, and his ponderous flat horns will make the observer gaze in awe. If he feels disturbed or intruded on, he will start swimming for the opposite shore. After he has disappeared into the woods, one may well exclaim, "What a magnificent beast!" The National Forests still harbor some 6,000 of these animals, mostly in Alaska, Wyoming, Minnesota, and Montana. This majestic king of the forest must be preserved.

The pronghorned antelope used to roam the western plains by the tens of thousands, but he has almost suffered the fate of the bison. There are now about 16,000 on the National Forests, mostly in Wyoming, Arizona, Montana, and California, and their numbers are increasing. They are fleet of foot and exceedingly graceful. Their insatiable curiosity helped to destroy them, for they will attempt to examine any unusual object whether it be a handkerchief tied to a ramrod or the foot of a reclining hunter waving above the grass. The young, which are mottled white and brown, are almost invisible as they lie in the sagebrush—but they are often killed by eagles, coyotes, and other predatory animals. Though antelope usually frequent grassy parks, they are more and more seeking the protection of the forest. They are essentially grazing animals, but in the woods will browse on tender leaves and shoots, showing an especial liking for willow.

Two very picturesque animals seldom seen by the ordinary traveller are the Rocky Mountain goat and the Bighorn sheep. They inhabit such inaccessible areas that only the most hardy mountain climbers and hunters ever get a look at them. The goat is found in Washington, Montana, and Idaho on the most exposed mountain crags and ridges above timber line. He is well adapted to this arctic climate for he is covered with a warm, woolly fur over which is a long, yellowish-white, shaggy covering of hair. He is a most expert climber, fast and sure-footed. His hoofs hold equally well on icy surfaces or on jagged rocks. There are said to be about 18,000 of these in the National Forests. The Rocky Mountain, or Bighorn, sheep live in a similar alpine habitat high in the mountains of Colorado, Idaho, and Wyoming. Their massive, curved horns are much-desired trophies. Their principal enemies are mountain lions, coyotes, lynx, wolves, and eagles, and their numbers are steadily decreasing. They are about as numerous as the mountain goats. The sheep browse on the young shoots and leaves of alpine plants, but seem to prefer the moss which grows abundantly in the alpine tundras.

Among the carnivorous mammals that are classed as predatory are the grizzly bear, which kills big game and cattle, the mountain lion or cougar, which kills horses, calves, and game animals, and the coyote or prairie wolf, which kills sheep and deer. The black and brown bears of which there are about 50,000 in the National Forests, are by no means as dangerous as any of the above. They are more or less omnivorous in their diet, eating fruits, berries, honey, roots, fish, frogs, reptiles, birds, mice and many other kinds of food. They are generally considered to be friendly and harmless to man, but they certainly can wreck a supply tent when they are looking for good things to eat.

The grizzly bear of the northern Rockies is a tremendous brute often weighing over 1,000 pounds. If so inclined, he can handle a man like a play-toy. He is America's fiercest and most dangerous big game. He once roamed quite generally through the western mountains, and years ago used to come down on the plains and feed on the bison. The light-colored tips of brownish hair give him a grizzly appearance, and the three-inch curved and steel-like hooks on his front paws assist greatly in upholding his dangerous reputation. The Forest Service does not want to exterminate him, but is inclined to relegate him to wilderness areas where he can do little harm. There are between 800 and 1,000 in the National Forests of the West.

The mountain lion is also a dangerous predator. He is a cat-like animal, often weighing from 250 to 300 pounds, which feeds largely on deer, though he may also feed on small mammals, birds, calves, young horses, moose, and elk. He can outrun a deer, but he more frequently stalks his prey, lying in wait on a rocky ledge or pouncing down from a tree. Anyone who has heard his scream in the dead of night—on the high mesas of western Colorado or in the manzanita brushfields of north California has something to remember him by. The cry is a combination of the thunderous roar of a lion followed by the bloodcurdling shriek of a woman in deadly terror. When groups of these animals get too destructive the Forest Service sends hunters to kill them.

The coyote or prairie wolf is a very fast, cunning, and intelligent animal which inhabits the open plains. He is also found on the open, sage-brush flatlands that occur in the mountains. Living in holes or caves, he feeds on jack rabbits, sheep, mice, ground squirrels, frogs, lizards, birds, and insects. The coyote looks like a wild dog, a sneaking, skulking creature, rarely seen in the daytime—but at night numbers of them are frequently heard in the timber and brush, near the lower timberline. The writer well remembers a certain oak thicket that had to be traversed every Sunday evening on the way back from a little mountain town in Colorado to the Ranger Station. It was a veritable bedlam of howling, whining, coyotes—thousands of them, one would say—but actually not over a half-a-dozen.

A full description of the various wild animals found throughout the forest regions of the country would require a large volume in itself. The few word pictures that have been given will perhaps indicate, among other things, the necessity of knowing the life histories, habits, food, shelter, and natural enemies of the various species before it becomes possible to establish a basis for their management and control.

Federal Objectives

No doubt the largest single factor which has helped partially to restore the forest cover for wild life has been the protection of forests from fire. The resulting reproduction of trees and brush that have sprung up during the last twenty-five to thirty years has been a great aid to wild life. Conservative lumbering as practiced by the Forest Service, the Indian Services, some state governments, and a few private individuals usually aids wild life, because part of the stand, namely the young growth, is left uncut and the forests are protected from fire. This protection, together with the reduction of hunting and the control of predators, has brought about notable improvement, especially when adjusted to the browse and other feed available.

That the management of our National Forests on a continuous-production basis over the last thirty-five years has materially aided wild life is shown by statistics. More than 70 per cent of the big game of the West is now in the National Forests. Census figures indicate that the big game population on these forests has increased 170 per cent from 1924 to 1938. The 70,000 miles of trout streams within the National Forests also show the result of years of careful fire protection and regulation.

Finally, so far as forest wild life goes, the Forest Service is aware of the perplexing questions of land utilization that arise on different areas within its jurisdiction. The sustained annual yield of big game is being correlated with other uses. The cattlemen want range, the hunter wants game; the silviculturist wants more forests and lumber and at the same time more young growth, while the biologist wants more primitive areas where wild life can thrive. These are not necessarily conflicting aims. They call for a judicious balancing of uses and sometimes the segregation of these objectives on different areas peculiarly suited to the different types of management.

The three main objectives of the Forest Service in wild-life management are: (1) to build up the habitat; (2) to increase wild-life population on some areas; and (3) to control the wild-life population so as to assure a balance. This can be done without jeopardizing grazing, forestry, or recreation interests.

It is one thing to gather scientific data and lay out a plan of action; it is quite another to acquaint the forest-using public with the facts. Intelligent public action on wild-life management must be based upon a knowledge of the established facts and principles and the cultivation of constructive mental attitudes and habits with reference to this resource. No conservation agencies, be they federal, state, or private, can hope to produce, control, and protect wild life unless there is created a healthy public attitude as a result of which every citizen assumes individual responsibility. Over 17,000,000 persons hunted and fished in the United States during 1936. But probably fully as important is the fact that the millions of citizens who visit recreational areas spent, according to Robert Marshall, about 250,000,000 man-days during 1931 in the pursuit of recreational enjoyment in the forests of this country. The number has since greatly increased. It is, of course, vital that hunters and fishermen do their part. I think the vast majority of them do. It is equally vital that the millions of other forest visitors use the forests in such a way that their activities will not interfere with the propagation of wild life. How to reach this vast audience is an educational problem of no small dimensions.

Progress in Wild-Life Restoration *

Wild-life restoration problems are so thoroughly interrelated with the restoration of forests and the perpetuation of watershed and recreational values that it is desirable to review briefly the diversified conservation work being done by the numerous public and private wild-life agencies. This work not only makes an interesting record of accomplishment, but it also emphasizes the need for a more coöperative and unified effort on the part of all the scattered organizations working in the field.†

About thirty-five years ago President Theodore Roosevelt set aside the first national wild-life refuge. Since then, and especially in the last nine years, the movement has received great impetus. Today over 280 areas totalling over 13,700,000 acres have been set aside for this purpose under the administration of the Federal Fish and Wildlife Service.

There are three kinds of refuges: (1) island refuges for birds that nest in colonies; (2) big-game refuges; and (3) migratory waterfowl refuges.

Island refuges for harboring colonies of nesting birds number fifty-one, and they protect pelicans, gulls, terns, herons, egrets, and similar birds. These areas are scattered along both the Atlantic and Pacific coasts and the Great Lakes region.

Big game refuges were started in 1908 with the establishment of the National Bison Range in Montana. On about 17,000 acres of land some 400 bison with small numbers of elk, mule deer, and mountain sheep are protected. Three other national refuges harbor herds of bison. They are the Wichita in Oklahoma, the Niobrara in Nebraska, and the Sully's Hill in North Dakota. Alaska also has a herd of some 200. Winter feeding ground for elk, now comprising some 18,000 acres of hay lands and winter range, have been established at Jackson Hole, Wyoming. Winter feed for the elk now in the area, some 20,000 animals, is still a problem.

The Wichita National Game Preserve of 61,000 acres is one of the most important upland-game research areas. Laboratories and a permanent research staff have been provided. The Hart Mountain

* Based on the following articles in *Field and Stream:* I. N. Gabrielson, "Our National Wildlife Refuges," Nov. and Dec., 1942; Harold Titus, "What Happens to Your Ammunition Tax?" July, 1942; "Ducks Unlimited," March, 1942; "The Truth about the Izaac Walton League," Jan., 1943; and "What Is the American Wildlife Institute?" Feb., 1943.

† See *The Status of Wildlife in the United States,* Report of the special committee on the Conservation of Wildlife Resources, Senate Report 1203, 1940.

Refuge in Oregon and the Charles Sheldon Refuge in Nevada, totalling over 800,000 acres, were established in 1936 for protecting the pronghorn antelope, of which there are about 7,000 head. Four large arid-upland game reservations have recently been added in Nevada and Arizona to protect mountain sheep. These total 3,600,000 acres. A number of refuges have been made possible by donations from the American Bison Society, the Izaac Walton League, the National Audubon Society, the Boone and Crockett Club, and other public-spirited wild-life organizations.

Other national game refuges are found in Maine, South Carolina, Florida, Mississippi, Georgia, Kentucky, and Wisconsin. For the most part these were sub-marginal farm lands purchased by the Rural Resettlement Administration and constitute a step in the establishment of a system of national upland game areas for the protection and feeding of many kinds of birds and mammals.

National migratory waterfowl refuges are concerned with feeding, breeding, and resting areas for geese, ducks, swans, cranes, and other kinds of waterfowl. In 1908 President Theodore Roosevelt set aside the Klamath Lake Waterfowl Refuge followed by the Malheur, the first of their kind. Their history is a good example of the start of a good movement without the public support. Without funds to patrol them, they soon were drained for agricultural development. Since they turned out to be of little use for farms, however, they soon reverted to barren alkali flats. After some twenty years of agitation these lands promise again to be flooded and restored to the waterfowl.

Up to about ten years ago our waterfowl habitats were in a bad way. The drainage of lakes and marshes, the fall of the watertable that resulted from deforestation, and the destructive action of floods and erosion had reduced our waterfowl areas to the danger point. In the face of this shrinkage of adequate feeding, nesting, and resting grounds, hunting increased by leaps and bounds, aided and abetted by the automobile, better roads, and better weapons. There were less fowl and many more hunters to go after them. Had it not been for a valiant band of conservationists, who saw what was happening and took steps to prevent it, our bird life would have suffered disaster.

The laws that were really the beginning of our present wild-life protection and restoration program were the Migratory Bird Law of 1913, the treaty with Great Britain in 1916, and the Migratory Bird Treaty Act of 1918.* These recognized the all-important fact

* A Migratory-Bird Treaty was ratified with Mexico in 1937.

that the problem of migratory birds is a continental problem and not a local one.

Our waterfowl, due to their migratory habits, require three distinct types of refuges. During the breeding season these birds are distributed through the northern and western states, Canada, and Alaska, while during the winter season they are found in narrow areas along the Atlantic and Gulf Coasts and the lower Mississippi Valley. During the spring and fall months they are *en route* between these two regions. Hence, breeding areas in the North, winter refuges in the South, and resting and feeding areas between these two regions are needed.

The program to start this system was launched in 1924 with the passage of the Upper Mississippi Refuge Act. This was followed by the Bear River Marsh Act of 1928 and the Migratory Bird Conservation Act of 1929. Refuges in Nebraska, Maryland, North Carolina, and Florida were among the first to be purchased. The restoration by the federal government of 7,500,000 acres of marsh lands in the United States for the use of our waterfowl is an objective that has already been more than half realized.

Many wintering-ground refuges have been acquired along the coasts of our southern states. The marshy areas in these states are ideally suited to this purpose. In Virginia, a 9,000-acre refuge is primarily valuable for Canada geese, redheads, canvasbacks, and snow geese. In North Carolina 88,000 acres in two areas help feed tens of thousands of swans, Canada geese, canvasback and other ducks. In South Carolina an area of 63,000 acres at Cape Romain is a wintering ground for pintails, black, and baldpate ducks. In Georgia, the Okefenokee swamp of over 300,000 acres provides a winter home for a great many species of waterfowl and other birds and mammals. Florida, Louisiana, Mississippi, and Texas have similar refuges.

Refuges of a second type provide resting and feeding grounds during migration, and hence they are located on the main flyways. The Upper Mississippi Flyway Refuge is probably the greatest of these, comprising 145,000 acres extending almost 300 miles along the river from Rock Island, Illinois, to Winona, Minnesota.

The third kind of waterfowl refuge is that needed for breeding grounds. More than 100 breeding-ground areas for waterfowl have been created in the northern states of Michigan, Wisconsin, Minnesota, and North Dakota, and in the western states of Utah, Oregon, Montana, and California. In upper Michigan the flooding of a marsh drained many years ago will provide 100,000 acres for all kinds of ducks and other forms of wild life. Large breeding-ground restoration of some 150,000 acres in North Dakota is the most important project of the kind yet undertaken by the federal government. Finally, the restoration of the swamps of the Malheur area—160,000 acres in Oregon—and the Klamath Basin must also be mentioned as contributing greatly to the restoration of the wild-life population of the Pacific Coast.

While the federal system of wild-life refuges is the most important, we must not overlook the fact that other federal, state, and private agencies have contributed largely to the restoration of wild life. Mention should be made of the Pittman-Robertson Act, Ducks Unlimited, the Izaac Walton League, and the American Wildlife Institute. The work of the states in fish and game restoration will be discussed in a later chapter.

Since 1932 a 10 per cent tax on sporting arms and ammunition has been collected by the federal government, but not until 1937 was this money definitely allocated by the Pittman-Robertson Act to be used for wild-life restoration. Up to the end of 1941, about $1,800,000 had been allocated for research, about $1,600,000 for actual development work, and about $1,300,000 for land acquisition. The money was allocated to the states for this work. Projects were devised and set up by the states and carried out under the supervision of the United States Fish and Wildlife Service, which sees that they conform to the spirit of the act.

At the present time, forty-six out of the forty-eight states have funds allocated to them. The conditions under which the money is granted are: (1) that the state agencies administer their own game affairs; (2) that enabling legislation has been passed in that state to avail itself of the terms of the act, the state appropriating twenty-five cents for every seventy-five cents allocated by the federal authorities; and (3) that no monies raised within the state by the sale of hunting licenses be diverted to any work not connected with game conservation. The amount of money given to each state depends upon the area involved and the number of hunting licenses sold in that state.

Up to June 30, 1941, 557 projects had been approved, 208 completed, and 60 were pending. Thousands of acres of game land have been acquired. Refuges, improved breeding areas, and rejuvenated game ranges have been developed. Pennsylvania acquired 70,000 acres of game land scattered in 70 parcels. New Mexico and Texas transplanted to unused ranges hundreds of pronghorn antelope. Utah has its Ogden Bay duck restoration project in which miles of dykes were built to make water available to waterfowl. Virgina made a wild turkey inventory, studied turkey range, and investigated feeding habits. These projects are typical of hundreds of others, and they indicate again the advantage of federal and state coöperation in conservation work, especially since most state wild-life restoration funds are inadequate to meet the needs.

Another good practical movement to save vanishing wild life is an organization called Ducks Unlimited, an outgrowth of the More Game Birds for America Foundation.

Western and central Canada has the greatest concentration of duck nesting grounds on the Continent. Most of the ducks killed in the United States are reared there. During the dry years in the early 1930's, when the water table dropped progressively lower year after year, wild ducks were hard pressed for nesting grounds. Not only drouth, but also swamp drainage, grass and forest fires, predators, and diseases took their toll. Under these conditions the movement to save the ducks was inaugurated. Since 1938 some 20,000 American duck hunters have been sending to Canada over $100,000 annually to help the work of building dams to create additional water areas for nesting grounds. Also grass-burning and damage by predators has been reduced. Areas so treated were set aside in perpetuity for the propagation of wild life by the provincial governments. Up to 1941 about a million acres of nesting areas, once stricken, had been rejuvenated. These comprised some seventy-five distinct projects.

One of our leading private conservation organizations is the Izaac Walton League of America. It was organized in 1922 to defend and conserve the woods, waters, and wild life of America. Great inroads in our wild-life population made themselves evident in the early 1920's, and the wild-life outlook seemed so dismal that about fifty far-sighted fishermen decided it was time to call a halt. They argued that since the selfish commercial exploiters of out-of-doors America were so strongly intrenched with lobbies in Washington to protect the "vested interests," it was high time that a strong, honest, unselfish agency should devote itself to the preservation of what was left of our wild-life habitats in a similar way.

The League's most sustained fight has been against stream pollution, but it has performed notable services in the solution of other wild-life problems. It was a powerful influence in the establishment of the Upper Mississippi Refuge, which was the first refuge to be established by Congressional appropriation. It purchased about 2,000 acres of winter range for the Jackson Hole elk herd and later turned it over to the government. It allied itself with the Quetico-Superior Council when the Minnesota and Canadian wilds were menaced by the hydro-electric power interests. It was instrumental in getting federal appropriations for the Bear River Marshes Refuge in Utah, and in general has stood behind the ten-year program for the federal acquisition of more migratory-bird refuges.

The American Wildlife Institute was started in 1935 to supplant the American Game Protective Association which was organized in 1911. It is supported by funds from its members, from business firms, and from generous conservationists. The Institute is dedicated to the study of wild-life problems based on sound scientific research. At the Institute's annual meetings, problems are discussed by technicians in the various fields of wild-life ecology. Specialists in the study of birds, fish, mammals, and other forms of wild life attempt to lay the foundations for better management. The Institute is coöperating with the land grant colleges of ten states by allotting funds each year to investigate many types of problems and helping to train a growing army of technicians in wild-life management. Some of the problems attacked were as follows: studies of the bob-white quail; growing of cedar for winter deer fodder; intensive waterfowl studies; the study of fish predators; restoration of the Atlantic salmon in Maine; improvement of bass fishing; and the study of the striped bass of the Atlantic Coast.

Finally, the North American Wildlife Conference was organized in 1936 as a federation of all organizations interested in wild life and conservation in general. It is the outgrowth of the work of J. N. Darling, former (March, 1934, to November, 1935) Chief of the U.S. Bureau of Biological Survey. In striving to get things done he was met either by bland indifference or by red tape. Politicians reminded him that public money was not available unless a strong group of voters camped on the doorsteps of Congress. After his resignation he organized such a pressure group. He enlisted women's clubs, garden clubs, camera clubs, Audubon societies, sportsmen's clubs, state federations of women, and many others. The second conference, held in 1937, had 800 representatives from clubs aggregating some 3,000,000 members. The Conference held in Washington in 1940 had over 1,000 representatives of nation-wide organizations. These conferences were held under the auspices of the National Wildlife Federation.

CHAPTER V

FORESTS AND THE WATER SUPPLY

It is not possible to grasp the full significance of the relation of forests to erosion and stream flow unless we understand certain fundamental relations that exist between forests and water supply. These relations are very complex, largely because the forest is a complex organic structure of many parts. The leaves, branches, trunks, and roots are only a very small part of this structure. To these must be added the undergrowth, the fallen leaves, the dead debris, the decayed spongy leaf mold, the top humus soil, and the inorganic subsoil. Bacteria, fungi, molds, and animal organisms also play an important rôle. The combined effect of these on the reception and final disposition of the rain and snow that is received absolutely determines how much of this precipitation will be held back, how much of it will run over the surface, and how much will be fed through subterranean channels to springs, mountain streams, the water table, and lakes and rivers.

Many are the interacting forces which come into play. Unravelling these complex forces and resolving them into a number of elementary principles is not only an interesting scientific problem, but it is also an excellent example of how science interprets complicated phenomena of nature. The forester has called upon biology, physics, chemistry, meteorology, geology, soil science, bacteriology, and other departments of knowledge to help explain these relations. The great mass of European data has been generally substantiated by American scientists. Scientific principles are the same the world over. As a final result the average forester has a pretty clear understanding of the reasons why the destruction of forests on mountain slopes visits its inexorable wrath on the valleys below.

Most of these forest-water relations can be summed up in the following simple scientific precepts:

Interception

Forests intercept portions of the precipitation by their leaves, twigs, and branches. Part of the moisture is evaporated from them and prevented from reaching the ground, and the rest runs down the branches and trunks. When the leaves and branches are thoroughly wet the mechanical effect is to delay the movement of water towards the forest floor and soil and thence into the mountain streams. The amount of water thus intercepted varies with the species of tree, the crown density, the age of the stand, and the character of the precipitation and wind movement. A great number of observations in Europe show that a spruce forest will intercept about 39 per cent and a broadleaf forest about 13 per cent of the rainfall. The amount is greatest in a middle-aged forest and smallest in a young forest. It is larger in light rains than in heavy ones.

The U.S. Forest Service has taken many measurements at paired stations inside and outside the forests of various types for several summers, in Maine, Massachusetts, Idaho, New Jersey, and Wisconsin, which show that forests intercept on the average about 25 per cent of the rainfall. This agrees quite closely with European data.

The interception of snow has received special attention because, usually, more of it will stick to the trees. In Europe a seventy-five-year-old stand of birch intercepted only 5 per cent, while a dense stand of spruce retained from 50 to 55 per cent. Similar measurements in dense forests of Idaho show that about 25 per cent of the snowfall on the average is caught by the crowns.

Evaporation

Forests reduce the melting and sublimation of snow under them and the evaporation of moisture from the forest floor and soil by preventing the wind and sunlight from reaching them. This not only conserves the snow but also holds back a considerable amount of water in the soil, with the result that the flow of streams, wells, and springs in near-by areas is greatly prolonged through the summer months. As nearly everyone knows, springs will often dry up after the forests around them have been removed.

The rapidity with which snow melts and sublimates under the forest cover varies with the species, density, age, and location of the stand. Snow disappears first from the clearings in the forest, next from the young stands, and then from birch, pine, and spruce forests in the order given. Ten years of observations in Russia showed that snow-melting lasted from three to seven weeks longer in the forests than in the openings, and remained longest under dense spruce forests. Similar observations have been made in this country. Investigations on the retardation of snow-melt in Idaho, Nevada, Arizona, and Washington show that snow disappeared from one to five weeks earlier in the open than under the forest cover.

A large part of the precipitation in our mountains comes in the form of snow, which in some of the high western mountains often lies until June and July. By delaying the melt of these snow remnants, the forests of the West perform an important service not only in reducing spring floods but also in storing irrigation water for use during the dry summer months.

The forest cover and the forest floor materially reduce the amount of evaporation from the forest soil at all seasons of the year. Lower air temperatures, lower soil temperatures, increased humidity of the air, and reduced wind movements hold evaporation in the forest to only a fraction of what it is outside. European data show that, on the average, evaporation from a forest soil with a leaf litter is only about 15 per cent of the amount evaporated from a similar soil in the open. Since under average conditions the evaporation for a bare soil is about 50 per cent of the annual precipitation, under the forest this would be cut down to only 7.5 per cent of the precipitation. Thus in a region that has a rainfall of forty inches a year, a bare soil would lose twenty inches of it in evaporation, while under a forest cover with leaf litter only three inches would be lost. This saving of water usually far exceeds the losses due to transpiration and evaporation from the trees.

Soil Packing

The crowns, branches, twigs, and leaves of the forest cover, by intercepting the rain, reduce its pelting force and thus prevent the soil from becoming hard and packed. As a result, surface run-off is reduced and absorption of water is increased. Most of us who have taken care of small garden plots are familiar with the packing force of a driving rain, especially if our soil is composed partly of clay. After a rainstorm it is necessary to cultivate the soil again, thus forming a mulch to check evaporation. In forest openings, especially where there are no dead leaves and humus, the soil is packed harder by each succeeding rain. If the slope is considerable, most of the rain will run off the surface and some of the soil will be taken along with it. Under the forest cover, the water falls more quietly and gently on a soil which is not packed but is covered more or less with leaf litter which makes it permeable to rain water.

Insulation

On account of the insulating effect of the forest cover and the forest floor, the soil is less likely to freeze than it is in the open fields, or at any rate it does not freeze so deeply nor for so long a time. This means that there is more absorption and less run-off in the forest, and hence there is less surface washing and more water is saved for the water table which feeds springs and streams.

In our northern states and elsewhere in the high mountains the ground is still frozen in open fields and clearings when the snow begins to melt in early spring. In a prolonged thaw the sun strikes the snow directly, especially on southern exposures, causing it to melt rapidly and run off in great quantities, even on the more gentle slopes. During the night freezing temperatures are apt to coat the open field with a glare of ice, over which snow water runs off as though from a gable roof when thawing resumes the next day. The greater the destruction of the forest cover by logging and fire, the greater is the run-off of water from these frozen surfaces.

Conditions under the forest cover are radically different. To begin with, the forest soil is warmer than the soil in the open, hence less likely to be frozen, because the trees prevent it from radiating its heat. Furthermore, the insulating effect of the forest litter, which itself is a poor conductor of heat, also prevents it from cooling off. Finally, the protecting blanket of snow under the trees also prevents the soil from freezing, as in agriculture, where it is well known that a snowy winter prevents crops from freezing.

For these reasons the soil under the forest either does not freeze at all, or freezes much later in winter, or to a much less extent than in open fields. It is very apt to thaw out earlier in spring, permitting underground seepage to feed the springs and streams, while open ground is still frozen hard. The unfrozen condition of the forest soil might even cause it to feed underground water slowly to streams all winter. All these hydrological conditions are well illustrated in our northern coniferous forests, especially in the spruce type in which the protecting cover is densest, the snow remains longest, and the absorbent litter is deepest.

Colloidal Binding

The forest cover and the forest floor protect the soil from physical deterioration due to excessive exposure to the sun and other weathering factors. This protection increases the absorbing power of the soil and consequently reduces run-off and erosion.

The opening of the forest cover by fire, excessive cutting, or other agencies exposes the soil to agencies which destroy its most characteristic physical properties. Such loss of physical structure is an invariable forerunner of wind and water erosion. Exposure to the heat of the sun dries up the organic matter of the soil with its important humic colloids. Wind and water remove these dried-up organic constituents and finally the inorganic soil beneath it. The depletion of these colloids causes the soil to lose the very organic ingredients which enable it to hold water and to hold the soil particles together, and which give it the fertility necessary to grow forest crops.

No other known soil constituents have the water-holding capacity which these insoluble humic colloids have. Nor can any other substance give soil its tilth, or loose, porous, crumb-like structure. These humic colloids have a gelatinous consistency when wet and are powdery when dry, thus binding the soil particles together when moist but allowing them to disperse freely when dry. This consequent breaking down of the crumb-like structure of the forest soil, enabling water and wind to tear loose the individual inorganic soil particles, is the beginning stage of all erosion. Imperceptible at first, the cumulative breaking down of this cohesive structure in humus forest soil soon develops into sheet and gulley erosion, primarily because of the decrease of water absorption and the increase of surface run-off.

Flocculation

Organic matter imparts a flaky structure to soil, quite independent of the binding action of the colloids just described. The organic matter of the forest soil, which is light in weight, completely changes the physical structure of the heavy inorganic soil by imparting a loose, porous, and crumb-like structure. This of course increases percolation and decreases surface run-off.

This is more true in impermeable soils like clay than in permeable soils like sand or gravel. Heavy, sticky clay is very resistant to percolation; but when mixed with humus it becomes loose and porous, permitting the water to soak in and reducing surface run-off. There is also a good effect in sandy or gravelly soils. Humus tends to bind these loose soils together and thereby prevents too rapid percolation to the water table. Thus underground seepage reaches the streams more slowly.

This action may be noted particularly in the gravelly soils of the Adirondacks and northern New England. During a heavy summer rain the gravelly soils, when devoid of humus, act like a sieve, letting the rain through nearly as fast as it falls. After the rain, they dry out rapidly. But under the spruce and hardwoods near by, where the same soils are mixed with humus, the surface layers retain large amounts of water. Often this water-retaining capacity extends down two feet or more and is noticeable for days.

Mechanical Resistance

Accumulated snow, litter, humus, dead branches, decaying stumps, and other debris on the forest floor offer mechanical obstruction and resistance to surface run-off. This results in increased absorption and percolation and in reduced run-off and erosion.

Aside from the absorptive power of snow and forest debris, this material also exerts considerable effect in delaying surface run-off by reducing the velocity and the carrying power of surface water. Every mechanical obstruction on a slope which can delay the raindrop's journey to the streams affords just that much more chance for the water to be absorbed or to find a more circuitous route to the rills and rivulets which join to form our large rivers. This has the ultimate effect of delaying the critical high-water stages of the large rivers and of giving valley populations more time to evacuate areas threatened with floods. Finally, erosion is greatly lessened, for experiments have shown that where the absorbing litter is removed the sediment carried away may be 1,000 times as great as on areas where the soils are protected by forest litter.

Absorption

The litter and humus of the forest soil are capable of absorbing water to an amount of three to nine times their own air-dry weight. This water storage, aside from other hydrological factors, is very important in the regulation of stream flow.

The spongy leaf mold that forms the natural forest floor will be found to vary in depth, depending upon the density of the forest, the species making up the stand, the frequency of forest fires, and the degree of cutting that has taken place. It is apt to be much shallower in open oak and pine forests than in spruce and fir stands, or in mixed conifers and mixed conifer-hardwood stands. In the spruce-hardwood type of New England it may reach a depth of from six to twelve inches; in the same type in the mountains of West Virginia and in the virgin hemlock-hardwood stands that once covered most of Pennsylvania, it may reach a depth of from twelve to eighteen inches. The humid and comparatively mild climate which prevails along the northwest coast is favorable for the formation of a very deep leaf mold, often as much as three feet. The deepest leaf mold the writer has ever seen was in the virgin spruce stands of the central Adirondacks, where in favorable localities the material ranged from three to five feet deep, representing the accumulation of centuries.

The amount of water retained by such a spongy surface layer of partly and fully decomposed leaves, twigs, and other dead debris is, of course, enormous. One European authority found that spruce-needle humus and beech-leaf humus retained 3.38 and 4.41 times their own air-dry weight of water respectively. Since the leaf mulch on an acre of virgin spruce forest weighs over six tons and that in a beech forest weighs four-and-a-half tons, air-dry weight, the litter of these forests is capable of retaining an average of about twenty tons of water per acre. Some European calculations place this figure as high as fifty tons per acre under favorable conditions.

American observations give similar results. The Forest Service and other agencies have found that coniferous forests in the southern Appalachians, hardwood types of the same locality, and the spruce-hardwood types of New England retained 3.4, 4.6, and 9.0 times their own dry-weight of water respectively.

The meaning of all this to stream flow is obvious. Often the difference between ordinary high-water stages of a river and disastrous floods is the last foot or two of rise that spells disaster when it breaks the levees and dikes. A forested watershed often holds back these last few feet of water in the manner just described.

Percolation

Obviously water that is not retained by absorption in the leaf mulch must either percolate into the subsoil or run off the surface. The forest cover, as a whole, provides ideal conditions for drainage through the subsurface soil layers, thereby reducing surface run-off to a minimum. Numerous decaying tree stumps, living and dead roots, rootlets and other vegetation, by leaving deep channels in the soil, facilitate the flow of water to the subsoil. The decay of this organic debris in the soil also promotes the activity of ground insects, worms, and other animals. These burrowing organisms honeycomb the soil with their tube-like cavities, thus keeping the subsoil loose and porous and greatly facilitating the underground flow of water. When the forest cover is ravaged these favorable drainage conditions are destroyed, and rapid surface run-off is usually followed by erosion.

Water Storage and Soil Anchorage

The forest, through its slow accumulation of organic matter, often over a period of a century or more, gradually increases the depth and the volume of the soil, thus creating greater reservoirs for holding water. This increase in soil volume is accomplished in two general ways: first, by the addition of leaves, twigs, and other forest debris which decay and form part of the soil, and second, by assisting the chemical and physical weathering of the underlying rock. The former process adds to the volume from above and the latter adds to it from below. As a result, the benefits derived from the forest as a conserver of water are cumulative, increasing very definitely as the stands reach maturity.

The intricate network of tree roots, including millions of fibrous rootlets, extending often many feet below the surface, cling to and anchor the soil particles and help to prevent the erosion of the soil. The problem of erosion is more fully discussed in the next chapter—but it must at once be clear on the basis of what has already been said about the water relations of the forest cover and the forest floor, that the control of erosion through the retardation of the run-off is one of the most important hydrological effects of the forest. In addition to the anchoring effect of the roots, all the factors just discussed in this section must be regarded as fundamental in explaining why forests greatly reduce erosion.

Transpiration and Use

Forests actually use large amounts of water to build up tissues, and the surplus water not needed is transpired in large quantities through their leaves. Obviously this water is not fed to streams.

European investigations show that a fully stocked beech stand 115 years old consumes 2,140 tons of water per acre per year. At the age of only 50 to 60 years, it consumes 1,026 tons and at 35 years of age only 321 tons. Broadleaf trees like elm, beech, and birch transpire from 25 to 43 per cent of the annual precipitation. Evergreen trees usually transpire less.

The amount of water available for stream flow is equal to the amount of the precipitation minus that lost by interception and evaporation from crowns, that lost by evaporation from the soil, that which runs off the surface, and that used by the trees or transpired. In spite of the fact that the forest cover consumes considerable quantities of water and loses some by evaporation and transpiration, the large savings made by the forest in reducing evaporation from the soil and in converting most of the run-off into underground seepage and storage, more than offsets these losses in mountain forests. The net result of balancing all the factors against precipitation, based both on European and American investigations, is that mountain forests are still the greatest existing conservers of precipitation for stream flow, wells, and springs. No other vegetative cover can compare with it. And the more nearly the forest approaches a fully stocked condition, the greater is the beneficial effect.

Total Effect Upon Stream Flow *

The following summary gives the effect of these factors on stream behavior, in forested as compared with non-forested watersheds:

1. Forests reduce and often entirely prevent the erosion of soil and thereby keep stream channels free to take care of stream run-off.

2. At the time of maximum high water the deforested watershed would carry off considerably more water than the forested one. Flood crests would increase.

3. At the time of drouth or minimum water flow the bed of the stream of the deforested watershed would carry comparatively little water or would actually dry up, while the stream of the wooded valley would still yield considerable water.

4. The forests would equalize the flow through the wet and dry periods of the year, often without diminishing the total yearly discharge.

5. In semi-arid regions a wooded watershed may consume a larger amount of water than the bare one and therefore diminish the total amount available to streams during the year. However, the total amount of water available to stream flow is not nearly as important as the manner of its distribution throughout the year and the reduced erosion exhibited by the forested watershed.

6. In times of low water a deforested watershed is likely to have insufficient water for streams, wells, and springs, with disastrous effects upon agriculture, wild life, recreation, and water-using industries.

* For a complete summary of the relation of forests and water, see *Forests and Water in the Light of Scientific Investigation,* Report of the National Waterways Commission (1912), pp. 203-302.

CHAPTER VI

HOLDING THE SOIL

AMERICA IS WASHING AWAY. The cancer of erosion is gnawing at America's vitals. We are suffering from a kind of mass astigmatism in allowing the fertile crop-producing topsoil to be carried seaward right under our very eyes. Rugged individualism is leading us to agricultural poverty, as it did the Chinese. Rich humus soil which has been forming since the last Ice Age is washing seaward at the rate of three billion tons a year. Rich topsoil which took centuries to form is allowed to disappear in a few months. Will American civilization be living, thriving, and happy two or three hundred years hence, or will it be of interest only to archaeologists and students of ancient history? In the year 2500 will men tell the story of American culture as it waxed, waned, and then perished, like the prehistoric cultures of Asia and Europe now being dug out of the sediments and debris of ancient rivers?

Such is the alarming picture presented in the latest literature on this subject.

The national government is in possession of full data on erosion, secured by surveys made under the National Resources Committee or under its predecessor the National Resources Board. An ex-chairman of the latter board put it this way: "...I believe that at our present rate of soil erosion, this country of ours has left to it less than a century of virile existence...."

This same agency points out that American prosperity has been achieved largely at the expense of capital loss of resources, the replacement of which may take a hundred years or more. We have used our forests, soils, wild life, and other resources with profligate extravagance. The result has been a large destruction of the very capital investment from which national income is derived. The rate at which our virgin vegetative cover has been stripped and converted into plowland and cut-over forest land probably has never before been equalled in the world's history.

What Is Erosion?

Erosion is usually defined as the removal of soil from the surface of the earth by natural physical forces such as wind, water, and glaciers. The first two are most important from our standpoint. Two kinds of erosion are distinguished: "normal" or "geological" erosion and "abnormal" or "man-made" erosion. Soil protected by natural vegetation undisturbed by man is eroded so very little that the effect is practically negligible. Foresters, geologists, and soil scientists are pretty well agreed that even this small amount would be more than balanced by soil-building processes. Abnormal erosion is caused by man's activities, such as lumbering, forest fires, wrong agricultural methods, over-grazing, industrial operations, and road building.

Geological erosion was at one time very much greater than it is now. It was probably greatest just after the retreat of the last ice sheet, which left large areas in the northern United States bare of any protective plant cover. During this period great wind storms piled up fine, wind-blown soils in the Mississippi Valley. These loessial soils are like very fine flour and are so easily eroded that their preservation is a great problem at the present time. After the ice retreated it probably took scores of years for the earth's surface to be again clothed with vegetation. In the great natural balancing process, that is nature's way of stabilizing the soil formed by weathering agencies. In this so-called "balance of nature" each climatic type is clothed with the type of vegetation best suited to the physical conditions. Thus grasslands are found in areas where rainfall averages less than twenty inches a year, deciduous trees are confined to the temperate humid areas, and evergreens largely to the cold, humid areas. The world over, steep, humid slopes are covered with forest growth.

Before Columbus and De Soto, some erosion no doubt occurred, as is indicated by the delta deposits of some of our rivers. Without man's interference many agencies probably contributed, among these being forest fires set by lightning, insect and fungus attacks which killed trees, violent storms, droughts, excessive rainfall in treeless regions such as the Missouri basin, landslides, avalanches, and other natural phenomena. As we have said, this erosion was probably localized on small areas and as a whole was balanced by new soil that was being formed in the forests from decayed leaves.

Finally—and this is most important—under regulated forest conditions such as obtain in most of western Europe, man is able to minimize if not entirely eliminate most of the conditions which have led to this so-called geological erosion. In these countries where land usage is generally in accordance with approved scientific methods and is properly balanced between forests, agriculture, and grazing, every acre is put to its best use and soil is considered capital investment that must be zealously safeguarded. Economic necessity brought about by dense populations demands this.

Man-made erosion is a different story! Since our forests are a mere vestige of those that once protected the soils of the United States, and since large areas of sloping land that never should have been plowed have nonetheless been cultivated for crops, abnormal or man-made erosion has become a serious national problem. Its ever-increasing progress makes it a deadly menace to our future welfare.

Factors Determining Erosion

The proper understanding of the erosion problem is dependent on the evaluation of the various factors that govern the rate of erosion. In general these are:

1. The nature of the vegetative cover.
2. The character of the underlying rock.
3. The character of the soil.
4. The velocity of the stream.
5. The load carried by the stream.
6. The climate, especially the amount and character of the rainfall.

The only one of these factors that can be appreciably controlled by man is the vegetative cover. And that is an important fact to remember when devising ways and means of erosion control.

A fully-stocked forest (a stand containing as many trees as it normally can for a given species, site, and age) naturally offers the best protection to a slope. A forest cut on the selection plan, that is, the removing of only those trees that have reached a certain diameter, usually offers fair protection. Much depends, however, upon how the logging is done. If the ground is badly torn up when the logs are skidded out and if the logging roads are steep and deeply cut by hauling devices, the bare soil is easily torn loose by water, and erosion gets a good start. When fire destroys all or a large part of the vegetative covering, as is so often the case after private logging, erosion over large areas follows as a natural consequence. The removal of trees down to two or three inches in diameter, as is sometimes done in taking out firewood, wood for charcoal, and wood for distillation, often leaves the soil practically bare of any vegetative cover.

The underlying rock is an important factor in erosion. Sedimentary rocks whose cementing materials are soluble or whose layers contain numerous cracks and joints are of course more erodable than hard, glassy, unstratified igneous and metamorphic rocks. Thus, shales, sandstones, and limestones are usually more friable than quartz, granite, quartzite, diabase, gneiss, and similar rocks.

It follows, therefore, that the soils derived from sedimentary rocks will wash away easily if they are fine, light, and deep, and less easily if coarse, heavy, and shallow. Glacial deposits usually erode fast. The fine, deep soil of the Appalachian region erodes more easily than the shallow, gravelly soils of the New England Mountain region.

The velocity of a stream depends largely upon the slope or gradient, its volume, its load, and the shape of its channel. In comparing erosion on two watersheds these are often important considerations. The steeper the slope, the greater the velocity, and, of course, the greater the erosive power of the stream if other factors are the same. As the volume of water increases, erosion increases. Water runs swiftest if clear, and the velocity is slowed down as the load of rock debris increases. The shape of the channel is often important because the velocity is retarded by the friction of the water against the bed and sides. Crooked channels and wide, uneven, stream beds cause friction and slow up the velocity. Deep, narrow, steep, and more or less straight gullies are therefore very favorable to a high rate of erosion. The load a stream carries helps to gouge out or erode the sides and bottom of the stream bed. The amount of silt, sand, gravel, and rock debris carried depends upon the velocity of the stream and all the factors which affect it. Excessive loads, however, will decrease the velocity, hence streams wear fastest when carrying only a partial load.

Lastly, the various factors that make up the climate have an important bearing on this subject. The amount of rainfall and its character and distribution are most important. Other things being equal, a region having heavy torrential rains, like our southern Appalachian Mountains, would suffer more than the Rocky Mountain region where rainfall is light and much of it comes in the form of snow. The Sierras and the Cascades of the West Coast have heavy precipitation but most of it comes in the form of snow.

In many localities wind is a great erosion factor. The great central dust bowl is only too well known. The removal of the protective grass cover in a region subjected to wind storms is just as disastrous as the removal of the forest cover in a region of heavy rainfall. Moving sand dunes are erosion problems, both along the coast and inland. Michigan and Indiana have areas along Lake Michigan in which wind-blown sand is a constant menace to farms and forests. Furthermore, there are many localities in Maine, Massachusetts, and other states where deep, light sands predominate. Wind erosion will sooner or later become a serious problem if the present forest cover is not improved.

Of all the erosion factors, slope plays a most important part. About 75 per cent of the land of the continental United States has a slope in excess of 2 per cent. However, erosion can take place on areas where the slope is less than 1 per cent. It is pretty well agreed that slopes of over 20 per cent belong to the forest. There is danger in cultivating slopes steeper than this if definite measures are not taken to prevent soil-washing. Slopes between 10 and 20 per cent can be used for pasture if precautionary measures are taken. Slopes of less than 10 per cent comprise the larger part of our agricultural soils, but even these must be protected by appropriate cultural methods. What has occurred quite generally in the United States in the past is that as level agricultural lands have become exhausted or inadequate, the cultivation has moved higher and higher to the steeper, more erodable slopes. Grazing and pasture have likewise moved to the higher slopes to displace the forest. Thus our whole agricultural system has been getting more and more out of gear with nature.* Some farms are where grass ought to be, and other farms are where forests ought to be. Soil exhaustion is menacing good agricultural lands just as forest depletion is ruining potential forest lands.

The whole matter of erosion can be summarized and its magnitude more readily visualized in terms of the physical laws which govern it. When slopes are steep, the soils deep, the underlying rock friable, rains heavy, and the soil surfaces more or less bare of vegetation, erosion by surface run-off may reach colossal proportions. Once the vegetation is removed, either wholly or partly, the most vital factor is the slope of surface which directly affects the erosive and the carrying power of running water. The erosive power of water varies as the square of its velocity, and the carrying power of water varies as the sixth power of its velocity. Thus, if you double the angle of a slope you quadruple the amount eroded, and if the velocity of the stream is doubled from, say, ten to twenty miles an hour, the boulders that can be moved increase from one-and-a-half tons to a hundred tons apiece. Thus we have the explanation of the great rock masses that are moved by a normally small stream in times of flood and the resultant damage done by such debris to bridges, railroads, and transportation systems.

* The abandonment of sub-marginal agricultural areas on the steeper slopes, and putting them in forests, promises to reverse this trend.

Erosion on Forest Land

It is impossible to give any accurate figures as to the relative amounts of erosion caused by bad farm practices, overgrazing, and forest destruction. No doubt more erosion is caused by the first two practices than by the last one. However, much farm and forage land that is seriously eroded is merely abused forest land. It should never have been taken out of forest production. Also about one-half of our forest land—342,000,000 acres—is being grazed by livestock, and much of it shows the effects of overgrazing. Hence such lands also are forestry problems. Most forest erosion is caused by logging and successive forest fires that have denuded the soil.

While farmland erosion is outside the province of this book, it is briefly mentioned here because of its great importance in the total erosion picture. This erosion has reached the proportions of a national tragedy. About 4,000,000 acres have been ruined by wind erosion in the "dust bowl," though measures are now being taken to check this damage. On another 56,000,000 acres, more than 25 per cent of the farmland is seriously affected by wind erosion. Water erosion has removed practically all the topsoil from 135,000,000 acres, and another 50,000,000 acres have from one-fourth to three-fourths of the topsoil removed in patchy areas. Much of this land has been abandoned and much of it is sub-marginal land inefficiently farmed. A large part of it should be returned to forest. In addition to this destruction of farmland, nearly 600,000,000 acres of western range lands show increasingly serious erosion.

Forest land erosion has been less spectacular and less dramatic, but nevertheless just as real and formidable. It is probably going on to a greater or less degree on most of the forest area classified as having a major or a moderate influence for watershed protection, because most of this area has been cut over or burned over. The amount of soil removed per acre of forest land is probably much less than that removed from farm and range lands, because usually some form of vegetation will be left on the forest land, or will quickly develop after denudation. Annual and perennial herbs, brush, and stump-sprouts often give some sort of protection, at least temporarily; but this is very inadequate compared to a good forest cover, because annuals and perennials do not have sufficient root systems to hold the soil on slopes, and in other respects do not give the soil enough protection.

Probably the best quantitative picture of forest land erosion that could be cited is the testimony given before the Joint Committee on

Forestry in 1940 by Mr. E. H. Clapp, Acting Chief of the U.S. Forest Service.* It must be borne in mind that erosion is intimately related to floods because of the debris that fills up stream channels and causes rivers to overflow their banks. This testimony covers both phases of the problem:

> The national significance of watershed protection in which forests must play a part is shown by annual average flood losses for the past 13 years exceeding $110,000,000 and a loss of life during the period of 1,265 people. The Mississippi flood of 1927 alone killed 246 people and did $300,000,000 of damage. Floods seem to be increasing in frequency and virulence with the destruction of the forest cover. In regions in which forests once held soil, 10 per cent of the area is seriously eroded and 40 per cent moderately eroded. That means soil deterioration on 485,000,000 acres which were formerly forests. Forests, of course, will be restored on a part of that area and help solve that problem.

The early stages of forest erosion are not conspicuous. The litter and humus are the first to go. The light-weight organic materials which largely compose the black surface-soil layers are the next to go, exposing tree roots, rocks, and the yellowish-brown subsoil. Large areas of this subsoil are soon exposed and the erosive process is accelerated, because the soil is less capable of absorbing rainfall and, since it lacks the organic binding material, is less capable of resisting erosion. Sheet erosion then concentrates on these exposed areas, and gully erosion begins and proceeds at an ever-increasing pace.

A trip up any of our eastern mountain streams would disclose these early stages of erosion in great abundance. Later stages are brought about by the movement of greater masses of water and by greater velocities, both of which factors mean the removal of larger amounts of erosive materials and greater size of materials carried. Any hunter, fisherman, forest guide, or lumberman could supply ample evidence. Muddy streams after severe rain storms; sand bars, shoals, and other deposits in river channels; and narrow river bottoms choked with rocks and boulders following each storm are common sights.

The gouging action of rock-laden streams is often seen. Banks are undermined, trees topple into the stream, and those left standing often have their bark entirely removed. Railroad embankments are gouged out, leaving the tracks and the ties dangling high in the air. One can find the abutments of an old bridge that has been washed

* *Forest Lands of the United States*, p. 1889.

away, or places where two or three successive bridges, each more substantial than the one before, have been destroyed by mountain torrents.

From the bottom of the valley we look up a mountain slope. Rock masses have accumulated at the foot of many of these slopes, especially where tributaries join the main stream. Where the land is less rocky, landslides and soil avalanches are often seen. As we look to the higher slopes, bare, rocky areas and rock ledges are exposed to view after the topsoil has been entirely washed away. On distant mountains conspicuous brown scars indicate bare soil, and such gashes into the bowels of the earth continually get larger (as the valley inhabitants can tell you) as each successive rainstorm breaks away the edges and tumbles the rich topsoil, sand, gravel, and boulders into the valley below.

The final stage of erosion is reached when these mountain streams wreak their vengeance on the valleys. Swift mountain streams with their huge loads of mud and rock debris become increasingly erosive as they near the lowlands, and as the velocity of the stream is slackened, the river begins to deposit its load. Rocks and boulders are dropped first because they are heaviest, then follow gravel, sand, and silt. Much of the latter is left on fertile farm lands, in river beds, and in various types of storage reservoirs .

Most tragic of all is the effect of these deposits on the disposition of flood waters. The filling of the stream beds with rock waste and sediment means that these channels cannot hold the large quantities of water which are delivered to them by the denuded slopes. Such building-up of the stream channel diverts the water to the adjacent lowlands causing widespread damage and suffering for many miles on both sides of the stream. Moreover, floods increase in frequency, and often a comparatively slight fall of rain will cause a flood due to the elevation of the river bed. Dredging operations in overloaded stream beds is a common sight. The detritus brought down by the mountain torrents is built up by steam shovels into two parallel dikes to make a channel for the flow of water. Such operations can bring only temporary relief.

The basic cure for forest-land erosion is reforestation. Forest rehabilitation on a wide scale must bring about a restocking of our potential forest land. Only fully-stocked stands give adequate protection. The benefits of these methods are naturally slow in developing. Hence these methods must be assisted by stone and log check-dams, docking and cribbing along streams, dredging the stream beds, terracing the slopes, and gully planting. These stream improvements are expensive. In this case, most certainly, an ounce of prevention is worth a pound of cure.

Erosion and Water Storage

It is difficult to overstate the seriousness of reservoir silting. There are in the United States over 10,000 dams and reservoirs with a construction cost of at least two billion dollars. In the last ten years the federal government alone has completed or placed under construction some thirty-five dams costing another billion dollars. The value of the associated and related engineering structures connected with these dams no doubt would exceed their original cost.

Reservoir silting, however, goes beyond the destruction of mere money value. The damage cannot be computed in terms of the original storage facilities because substitute storage facilities cannot be developed at the same cost. For many reservoirs there are no substitute sites. Usually the best sites are chosen first; new sites can be utilized only at ever-increasing cost. Eventually, new sites cannot be found at all. Here we have the most effective argument ever advanced for the so-called vegetal method of erosion control which attempts to control erosion at its source.

Sediment reconnaissance measurements made by the Soil Conservation Service and other agencies on about 600 reservoirs show that silting is directly traceable to soil erosion. The speed of silting varies directly as the vulnerability of the watershed to water erosion. All kinds of reservoirs are losing storage capacity. The more recently built reservoirs designed for flood control have of course not yet shown appreciable silting, but their ultimate destruction is inevitable. Sooner or later all reservoirs will lose their usefulness if the present rate of erosion and silting continues.

Recent studies indicate that of the existing reservoirs of the United States:

38 per cent have a useful life of from 1 to 50 years.
24 per cent have a useful life of from 50 to 100 years.
21 per cent have a useful life of from 100 to 200 years.
17 per cent have a useful life of over 200 years.

This means that practically two out of every three reservoirs will be filled up with silt within 100 years if the present rate of erosion is not slowed up! Some extreme cases have come to light in the southern Piedmont Region where thirteen major reservoirs with dams averaging thirty feet in height have completely filled up, and their useful life was less than thirty years.

Erosion and Domestic Water Supply

The previous chapter on fundamental forest-water relations clearly shows the vital relation between the forest and our domestic water supply. Let us look at the economic phases of this important problem.

Water is one of the essentials of life. Cities, towns, villages, and rural areas must have an adequate supply. This is derived from lakes, streams, springs, and wells. Forests play an important part in supplying this need by protecting lakes, streams, and reservoirs from erosion deposits and from excessive turbidity, by furnishing a more equable flow throughout the year, and lastly by converting surface run-off into underground water, thereby maintaining the level of the water table which feeds wells, springs, and mountain streams. Regardless of whether domestic water is supplied from surface water or from ground water, forests play a conspicuous part in maintaining adequate supplies of clear, potable water at all seasons.

The 1934 report of the National Resources Board states that about one-third of the people in the United States live in ninety cities of more than 100,000 population, another third in smaller towns and villages, and the rest on farms and in rural areas. In general the urban populations depend largely on surface water (lakes and streams), while people living in the smaller towns and villages and on farms depend mostly on ground water (springs and wells).

Since storage facilities are not needed where the source of water is a lake or a large river, reservoir silting becomes a problem only in those cities and towns that must collect their water from numerous mountain streams and store it for future use. As cities grow, the need for water increases and the cities are forced to go farther back into the "hinterland" for their water, at ever-increasing expense. For the large cities located on lakes and rivers, the sewage disposal problem complicates matters. During the serious drouths of 1930 and 1934, adequate, uncontaminated water supplies were a very serious problem for many cities.

To get a better idea of the magnitude of the domestic water problem and the enormous investments made to bring it to the consumer, let us take a few well-known cities as examples and examine their waterworks establishments. This no doubt will bring home to us the seriousness of the reservoir-silting problem.

About 100 years ago New York City made its first bid for water outside its own boundaries. Old Croton Dam was finished in 1842. Soon the need for more water compelled the building of the Catskill and Schoharie reservoirs about 125 miles away, which increased the

Photo by C. T. Bodwell, copyright 1940

ECHO LAKE, WHITE MOUNTAINS, NEW HAMPSHIRE

Thousands of acres have been set aside for recreational purposes in New Hampshire by the state and federal governments.

SENTINEL PINE BRIDGE, THE FLUME, NEW HAMPSHIRE

In the Franconia Notch Reservation.

Photo by C. T. Bodwell, copyright 1940

Photo by C. T. Bodwell, copyright 19

THE FLUME, FRANCONIA NOTCH, NEW HAMPSHIRE

This beautiful gorge, with granite walls seventy feet high, is owned jointly by the state of New Hampshire and the Society for the Protection of New Hampshire Forests.

Photo by C. T. Bodwell, copyright 1934

OLD MAN OF THE MOUNTAINS

Rising 1,200 feet above Profile Lake in the White Mountains of New Hampshire, this rugged formation is said to have suggested to Hawthorne his story of "The Great Stone Face."

U. S. Forest Service Photo

TUCKERMAN'S RAVINE, WHITE MOUNTAIN NATIONAL FOREST

This famous resort for winter sportsmen is often filled with snow until late June.

DEER PARK SKI AREA, OLYMPIC NATIONAL PARK, WASHINGTON

U. S. Forest Service Photo

U. S. Forest Service Photo

IN HOLY CROSS NATIONAL FOREST, COLORADO

WOUNDED MAN LAKE, CUSTER NATIONAL FOREST, MONTANA
The hardy camper who is willing to rough it enjoys such remote spots.

U. S. Forest Service Photo

U. S. Forest Service Photo

IN THE SELWAY-BITTERROOT WILDERNESS AREA, MONTANA
The big-game population has more than doubled since 1924.

SAWTOOTH NATIONAL FOREST, IDAHO
Such recreational use of the national forests is increasing rapidly.

U. S. Forest Service Photo

U. S. Forest Service Photo

MIRROR LAKE CREEK, GALLATIN NATIONAL FOREST, MONTANA
This is in the Spanish Peaks Wild Area.

BLUE PARADISE LAKES, GALLATIN NATIONAL FOREST, MONTANA
Snow banks are still visible in August.

U. S. Forest Service Photo

U. S. Forest Service Photo

MAZAMA DOME, MT. BAKER NATIONAL FOREST, WASHINGTON

U. S. Forest Service Photo

CATHEDRAL PARK, NEVADA NATIONAL FOREST

U. S. Forest Service Phot

VIEW OF MT. BAKER, WASHINGTON STATE
This peak reaches a height of 10,750 feet.

U. S. Forest Service Photo

TRAIL TO CHARLESTON PEAK, NEVADA NATIONAL FOREST

There are over 150,000 miles of riding and hiking trails in the national forests.

Photo by Bayard Wootte

UPPER SOUTH: ENTERING NANTAHALA GORGE, NORTH CAROLINA
The Great Smoky Mountains National Park is just north of this scene.

LOWER SOUTH: CABBAGE PALMS, FLORIDA

U. S. Forest Service Phot

U. S. Forest Service Photo

NEAR MT. TOXAWAY, BLUE RIDGE MOUNTAINS, NORTH CAROLINA

LITTLE TENNESSEE RIVER AND GREAT SMOKY MOUNTAINS

Photo by Bayard Wootten

U. S. Forest Service Photo

TRAIL IN NANTAHALA NATIONAL FOREST, NORTH CAROLINA

U. S. Forest Service Photo

MOOSE IN LOLO NATIONAL FOREST, IDAHO

The Forest Service maintains scientific management of wild life in the West. Game as a national resource is evaluated at over a billion dollars a year.

CARIBOU IN TERRITORY ADJACENT TO JASPER NATIONAL PARK, ALBERTA, CANADA

Photo Canadian Travel Bureau

Photo Canadian Travel Burea[u]

BUFFALO OR BISON, BUFFALO NATIONAL PARK, CANADA

This animal became almost extinct before measures were taken for its protection.

BLACK BEAR AND CUBS, MT. BAKER NATIONAL FOREST, WASHINGTON

U. S. Forest Service Pho[to]

daily capacity to about 1,200 million gallons. By 1935 even this was seen to be inadequate, since consumption in the metropolis had increased to 1,157 million gallons a day. New reservoirs are now under construction in the upper Delaware watershed at a cost of about $275,000,000. And it is said that if the population increase is maintained, this supply will be inadequate by 1955, when new sources will have to be found. The total investment at the completion of the Delaware project will be in excess of $800,000,000. The protection of these watersheds is, of course, essential to prevent the silting of this expensive establishment. By 1938 the state had accumulated about 230,000 acres of forest lands in the Catskills. This is administered as the Catskill Forest Preserve. In addition, over 4,000,000 trees have been planted on the lands adjoining the reservoirs.

As early as 1652 Boston had its first public water supply system. As the city grew, more and more water was needed. Some years ago the large Framingham reservoir, twenty-five miles from Boston, seemed adequate for all needs. It was soon necessary, however, to provide the Wachusett reservoir, forty-five miles away. The city is now damming up tributaries of the Connecticut about 100 miles away to meet the ever-growing need for water in metropolitan Boston. The Framingham and Wachusett reservoirs have been well protected by forest plantations.

Boston and New York are indeed fortunate in being located in regions of adequate rainfall. Cities located in the semi-arid West often obtain their water with great difficulty and at great expense. Los Angeles has gone great distances and has made large investments for water. The original source of water supply, the Los Angeles River, soon proved inadequate and the city tapped the Owens River about 240 miles away. With a population of about one million at the present time, the city is growing so rapidly that new supplies must be sought. Already construction has begun on an aqueduct from the Colorado River 270 miles away, but even this will be later augmented by Mono Lake at a much greater distance from the city. The water systems by which southern California plans to solve its water problems, primarily for domestic purposes, will eventually cost in excess of $350,000,000.

San Francisco and environs, similarly situated in a region of inadequate rainfall, is obtaining a part of its public water supplies from Hetch Hetchy reservoir 200 miles away, at a cost of over $125,000,000. In both these projects the National Forests play a major rôle in protecting watersheds.

As a matter of fact, the administration and protection of the National Forests of the West is closely linked up with the domestic water supply problem of that entire region. About 50 cities with a population of 10,000 or over, and a total population of over 5,000,000, and 327 cities with less than 10,000 population and a total population of over 700,000 depend upon National Forest watersheds for their domestic water supply. The investment of these cities in their water works establishments is about $500,000,000. Furthermore, some of these have made such large investments largely because of the forest policy of the federal government which insures maximum protection.

As to the silting of domestic water reservoirs, many parts of the United States report serious damage. In the Southeast the rate of silting is high because of deep soils and sudden heavy showers. In the Southwest the rate of silting is high because of overgrazing. In southern California the rates are high because fires are allowed to consume the protective brush cover on the steep drainage basins.

A few typical cases will give an idea of the nature of the damage. The city of Spartanburg, South Carolina, built a city reservoir at a cost of $470,000 which lost 17 per cent of its storage capacity in eight years. At that rate it will be filled in less than forty years. A city reservoir at Waco, Texas, costing $2,000,000, lost 20 per cent of its storage capacity in six years and will be rendered useless in another twenty-four years if the rate of erosion is maintained. Another large reservoir supplying Decatur, Illinois, lost 14 per cent of its storage capacity in fourteen years.

Erosion and Hydro-Electric Power

Where the chief purpose of reservoirs is hydro-electric power development, we again come to a situation where enormous investments of money are involved. Practically every state in the Union utilizes power from its streams. The maintenance of a high standard of living both in the rural and urban areas requires cheap electric power. As our natural fuels become less available we will have to depend more and more on hydro-electric power for our energy.

According to the annual report of the Department of the Interior, the total capacity of all our water power plants on January 1, 1936, was over 16,000,000 horse power. Since the potential capacity of all water power sites in the United States is estimated at 80,000,000 horse power, it is plain that we have developed only about 20 per cent of our national water-power resources.

As to the part that water power plays in the total power developed by all methods, it might be said that in 1934 a little more than one-third of all power was developed by streams. That hydro-electric power is only in the infancy of its development is certain. While water power will have to compete with methods of developing power by fuel combustion, there is no doubt that water power can meet the challenge of this competition in many sections of the country.

The development of long-distance electrical transmission has aided water power development greatly. However, even though it is possible to transmit electricity from plants of large capacity a distance of from 250 to 300 miles, it is still true that the energy must be consumed relatively near the point of production. This is often impossible because many good water-power sites are at some distance from the dense population centers which consume the power. Coal, oil, and other fuels have the advantage that they can be transported to the center of consumption and then converted into energy. Hence, distance will be a limiting factor in hydro-electric development until improved methods of transmission can be invented.

The primary requirements of a good water-power site are a stream of uniform flow which at the same time is free from silt. The primary purpose of the dam is to give a sufficient head of water to operate the wheel or turbine. In streams of uneven flow the dam is also used for the storage of water against seasons of low water. The dam must be so built that it can withstand floods, ice, and erosion debris.

Variable stream flow is one of the big problems of the industry. Market conditions require a more or less uniform flow of power. Streams with pronounced high and low water stages cannot meet this situation unless steam auxiliary plants are used. These steam plants are often not feasible. The modern super-power system seems to solve this problem in the most economical way. The power plants (both hydro-electric and steam) of a large region are connected in such a way that power can be transferred from plants where there is a surplus to markets where there is a peak load due to a concentration of mills and factories. A very flexible system of distribution has been worked out, for instance, in the northeastern power zone, resulting in a great saving of fuel and equipment.

However, the reservoir, the uniform stream flow, and the silt-free water are still the keystones in the arch of water-power production and we know that the forest-covered watershed can contribute much towards furnishing these requisites and thereby insuring the life of the water-power industry.

Many water-power dams and reservoirs have already lost all effectiveness and many more have suffered a considerable loss of storage capacity. Our most conspicuous examples of reservoirs destroyed by silt are of this type, because water-power reservoirs were among the first to be built in this country. In the Piedmont Region of the Southeast there are numerous silt-filled storage basins with adjoining powerhouses and mills abandoned. In the western states the construction of auxiliary sediment-detention basins has helped to pro-

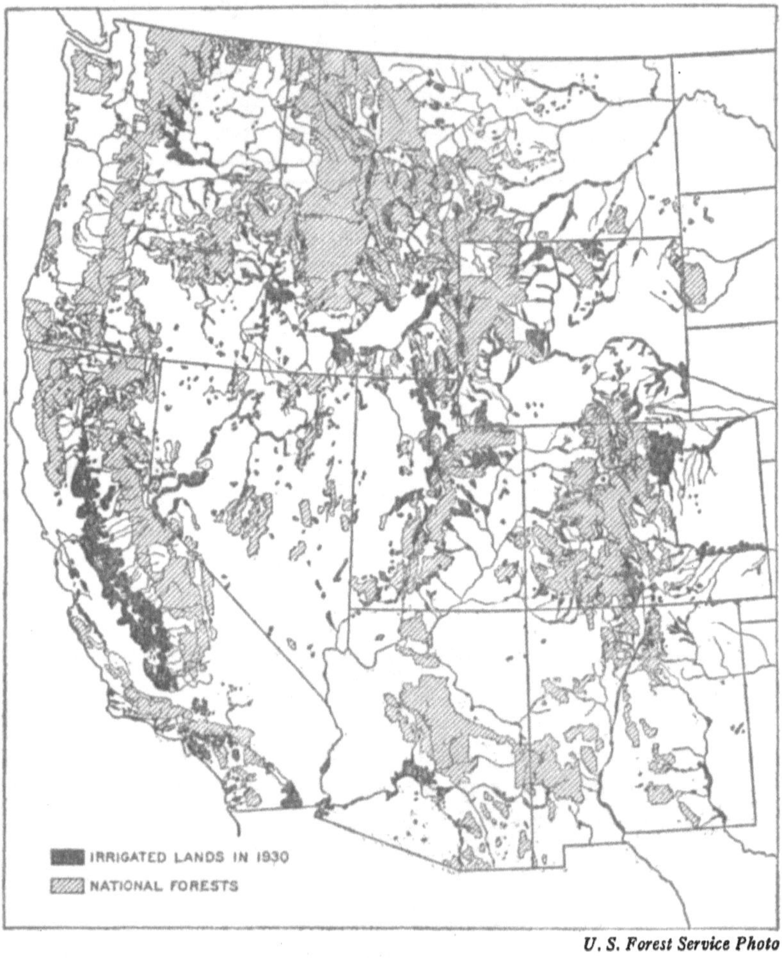

U. S. Forest Service Photo

Virtually every major stream west of the Great Plains, and most of the minor ones, head in a National Forest and supply a large part of the water for the great irrigated land areas

long the life of many plants. Certain reservoirs in southwestern Virginia, built for hydro-electric power, have been subject to high rates of silting, varying from 2 to 7 per cent of their capacity per year. One of the largest channel-type reservoirs in the country on the White River in Missouri, which provides power over a wide area, has lost 46 per cent of its storage capacity in twenty-two-and-a-half years.

The largest federal dams recently completed or in process of construction, in which water-power generation is one of the main objectives, include the Wilson Dam in Alabama, the Norris in Tennessee, the Wheeler in Alabama, the Grand Coulee in Washington, the Boulder in Arizona, the Kennett in California, the Bonneville in Washington, and the Fort Peck in Montana. These eight super-dams alone represent an investment of over $600,000,000, and there are many smaller ones besides. The rehabilitation and protection of the forests and the restoration of the grasslands on the watersheds of these streams must be considered a prime requisite to the long life of these enormous engineering structures, or otherwise their life will be short.

Erosion and Irrigation Water

Irrigation water is the life-blood of agriculture in nineteen states west of the Mississippi. According to the 1930 census almost 20,000,000 acres of land are being irrigated, and the value of the lands, buildings, machinery, reservoirs, and distributing systems reaches the enormous total of six billions of dollars. The crops were valued at $900,000,000 in 1929. After the present federal irrigation projects are completed this total will rise substantially. The restoration of the forest cover contiguous to these many reservoirs means conserving snow for late summer irrigation, distributing the irrigation water more evenly throughout an almost rainless summer, and keeping the reservoirs and ditches free of silt. The restoration of the grass cover on western grazing lands would likewise greatly lessen the silting of these valuable irrigation basins.

Many examples of ruined irrigation reservoirs could be cited. The Elephant Butte Reservoir, costing $5,000,000, built on the Rio Grande in New Mexico in 1915 to provide irrigation water, had lost 14 per cent of its storage capacity in twenty years. The Arrowrock Dam in Idaho, constructed the same year, had by 1930 lost 25 per cent of its capacity. Silt also clogged the canals and irrigation works. Even the enormous Boulder Dam with Lake Mead, the largest project of its kind in the world, costing over $125,000,000, will be comparatively short-lived if the amount of silt carried by the Colorado River cannot be materially reduced.

The National Forests of the West play a vital part in protecting the irrigation and water-power reservoirs, just as they safeguard the domestic water supplies of the region. We must remember also the closely affiliated industries such as livestock, lumbering, mining, transportation, recreation, and wild life, all of which more or less directly depend upon a good water supply. There are over 1,300 irrigation projects and over 300 water-power plants enjoying the protection afforded by these Forests.

Erosion and Flood Control

The danger of erosion also threatens the very engineering works designed to control floods. Erosion not only increases the height and destruction of floods, but it also threatens to destroy the down-stream engineering structures in a few generations, either by silting up the flood-control reservoirs or by filling up the stream channels, thereby making necessary the raising of the levees. There are extreme cases like the Austin, Texas, dam whose reservoir was completely filled in fifteen years, and there are large numbers where the average yearly silting amounts to from 2 to 5 per cent of their storage capacity.

The large modern flood-control storage reservoirs are comparatively new and as yet show little silting. There is no question that eventually they will lose much of their effectiveness in this manner unless adequate erosion control measures are taken. Of some thirty-six federal dams either completed or in the process of construction during the last ten years, about half are designed for flood control, though most of them have other uses also. The spending of hundreds of millions of dollars for this method of flood control cannot be justified unless there is a more concerted effort to prevent erosion.

The silting of stream channels is a very common occurrence in many parts of the United States. In many cases it seriously interferes with navigation, often preventing it altogether. We have already referred to the fact that this silting diverts the flood waters over the adjacent mainlands. Enormous sums are constantly being spent to dredge out these channels and keep them open for navigation. The Rio Grande in New Mexico, the Anacostia in the District of Columbia, the Galena in Illinois, the Arkansas, the Appalachicola, the Tennessee, the lower Brazos, the Colorado in the West, and the Colorado in Texas offer specific instances of serious sedimentation. The bed of the Rio Grande has risen five feet in nine years and that of the Arkansas in places five feet in the last forty-five years.

In the last twenty-five years especially, huge sums have been appropriated by Congress for the maintenance and improvement of rivers and harbors. The grand total for this kind of work now approaches two billions of dollars. It is very evident that we are in this case treating the symptoms and not the causes.

CHAPTER VII

FORESTS AND FLOOD CONTROL

IN THE TWO preceding chapters we have examined the basic forest-water relations and how these control run-off and erosion. In the present chapter we will find out how these same factors affect stream flow, and more especially, floods. Flood control is one of the great internal problems confronting this country today, an important part of the larger problems of national defense and security. Every high school student should know the facts and learn about the measures that are being applied. The study of this problem also shows most clearly that history repeats itself and that the solutions applied by older countries who have learned their lesson by bitter experience can and must be applied to our own conditions.

The development of forestry through the centuries shows that man has emphasized successively the wild-life, the timber, and finally the water phases of forest management. For a period of about five centuries forestry practice in Europe was based as much on the demands of the chase as on the need for wood products. The destruction of mountain forests focused attention upon the forest as a regulator of water supply. The reforestation of the watersheds of mountain torrents was not recognized as a proper public measure in German Austria until 1788. But the argument of the protective influence of the forest cover did not enter legislation until about the middle of the nineteenth century. Hence we can say that the idea of the forests as a protection from erosion and floods is only about a hundred years old. It was undoubtedly born of stern necessity.

The experience of France can be used further to establish this fact. As early as the sixteenth century decrees were enacted with regard to the cutting of forests in mountain districts. During and after the French Revolution these beneficial restrictions vanished. By 1803 complaints were heard throughout eastern France that unrestricted cuttings in the French Alps were turning slow-flowing mountain streams into dangerous torrents. As a direct result of this destructive cutting over 8,000,000 acres of tillable land had been rendered more or less useless and the population of eighteen departments had been impoverished or reduced by emigration within a few decades. A young French engineer, Surell, was the first to study the possible control of this evil and proved in his *Etude sur les Torrents,* in 1841, the relation of this problem to the forest cover and the necessity of controlling the torrents at their source. Even with the facts established, it was not until the Reboisement Laws of 1882 that the problem was seriously attacked.

In Europe, scientific forestry research (which there as everywhere else, paradoxically enough, has followed disaster, instead of preceding it) has methodically analyzed the whole forest-water economy and has corroborated, item by item, what experience had been teaching for more than a hundred years.

With this experience and research to fortify their arguments, American foresters early in our forestry history sought to make this forest-water relationship a primary reason for establishing National and State Forests. Most of our later organic laws looking toward forest conservation are based on this fundamental relation.

In this country during the last twenty-five years there is a growing perception of the fact that floods are becoming more severe, more frequent, and more destructive. The flood disasters of 1913, 1927, 1929, 1934, 1935, 1936, 1937, 1938, and 1943 are only too well branded into the public consciousness. The records show also that ever since the creation of the Mississippi River Commission, sixty years ago, the height of the flood levees has been constantly raised. Other rivers tell a similar story. In a number of localities, painted white lines on low-lying buildings, or cross-timbers nailed on poles or trees are used to mark the height of each successive flood. The trend of these flood-crest reminders is generally upward.

If more evidence is needed that floods are getting worse and have been getting worse for the last hundred or more years, let us take a look at the river deposits themselves. The geological history of Mississippi alluvial deposits shows conclusively that recent sediments laid down by the river since the agricultural occupation of the valley are not only much coarser in texture but far more diverse as to texture and color than those of the pre-agricultural period. Old deposits show a much finer texture and more uniform composition. The more recent deposits were laid down by waters very much more violent and waters that had much greater carrying power than those which laid down the older materials underneath. Furthermore, the depth of the newer, coarser, sediments is often much greater than that of the older and finer underlying sediments. Yes, man's occupation of the Mississippi Valley has terribly hastened erosion and increased run-off and floods. There is not the slightest doubt, because the river sediments say so. De Soto, Marquette, and La Salle saw the Mississippi in flood, but it was not a white man's flood—it was just a spring freshet—for the valley as a whole was protected by a natural plant cover.

Looking at the country as a whole, we must remind ourselves that seven-eighths of our original virgin forests are gone. With them has gone a large part of their enormous storage capacity for water. Is there any mystery in the fact that floods have increased in severity after the removal of millions of tons of water-absorbing leaf-mold? Or is there any mystery about the fact that fully one-fourth of our total land area presents watershed problems that are crying for urgent solution?

Let us look, however, at the more encouraging side of the picture. The year 1936 will no doubt go down in conservation history as the year of our real awakening. The full weight of the forester's arguments finally bore fruit in the form of a combined forestry-agricultural-engineering attack upon the problem. The Flood Control Act of that year brought together for the first time on a large scale the advocates of "up-stream vegetal control" with those of "down-stream engineering control."

In a problem of such magnitude, a further analysis is necessary. It will be worth while to survey the damage done by floods, the factors affecting floods, the general methods used to control floods, and a brief history of flood control work done down to the present-day flood control program.

Losses Due to Floods *

The average annual loss due to floods is estimated at between twenty-five and fifty million dollars. Individual floods, however, may take a toll of hundreds of millions within a few days. Many states are confronted with yearly losses on their smaller rivers which do not get the attention they deserve. For example, along the Illinois River the losses are said to average about two million dollars a year. We are attracted by the newspaper headlines of the great floods, but the many smaller ones escape our notice. Yet their aggregate damage is much greater than that of the few large floods, and over a period of years will run into the hundreds of millions of dollars. To give

* *The Land in Flood Control,* U. S. Department of Agriculture, Misc. Pub. 331, 1938.

even a partial list of floods would be futile, for the index of floods in the New York Public Library, mentioning only places and dates, requires over 700 pages.

If we wish to get a complete picture of the total cost of floods, we should add the government expense in attempting to control them. The sums spent by the federal government on existing river and harbor projects directly connected with stream flow (such as building levees, dredging, etc.) during the last fifty years will amount to a total of about two billion dollars. To this must be added salvage and relief costs of city, state, and federal governments, together with those of numerous private agencies like the American Red Cross. Finally, for death, injury, disease, loss of wages due to the paralysis of industries, injury to health, homelessness, and privation there can be no adequate estimate of cost.

The losses during the last twenty-five years have been greater than during any similar period in our history. In 1913 floods in Ohio killed over 500 people and the property damage exceeded $150,000,000. In 1927 the Mississippi had its worst flood on record. Official Red Cross accounts describe it as the greatest disaster this country ever suffered. It inundated 26,000 square miles, killed over 200 people, rendered 700,000 homeless, and effected property damage estimated by the War Department at over $300,000,000. Also in 1927 the Connecticut and other rivers in New England routed over 30,000 people out of their homes and wrecked entire villages in Vermont and Massachusetts. In 1929 serious floods visited the southern states, inundating about 1,500,000 acres. In 1933 and 1934 Washington, Idaho, and California experienced disastrous floods. In 1936 the Ohio River and the New England streams again went on a rampage, affecting seventeen states in all. The total damage exceeded $500,000,000, and there were 168 people killed, over 6,800 injured, and about 430,000 rendered homeless. The great flood in the Ohio Valley in January, 1937, exceeded all previous high-water stages on record from Point Pleasant, West Virginia, to Cairo, Illinois. At some places the record was broken by as much as ten feet, and the high-water mark reached the highest point previously estimated as possible. From the standpoint of property loss the 1937 Ohio Valley flood was by far the most disastrous ever experienced in the United States. And so the story goes, with worse floods in prospect.

Factors Affecting Floods

It has frequently been said that floods are not moving toward man, but that man is moving toward the floods. This is due to the many favorable conditions found in the valleys along the big rivers, all over the world. In the historical development of this country, as well as most others, the agricultural population settled in the rich valley bottomlands. Cities with their huge commercial and industrial enterprises were established along the rivers, which were also the first ready means of transportation. Roads, railroads, and communication facilities naturally followed the level and easy routes along the rivers. And of course the greater the use of these low-lying lands, the more valuable they became and the greater their liability to property damage by floods.

It has been suggested that centers of population be moved to higher ground, safe from flood waters. Man, however, refuses to abandon the valleys. Since he prefers to meet the challenge, he must turn to a study of floods and attempt to curb them. He finds that four physical conditions determine the extent and behavior of floods:

1. Precipitation.
2. Physiographic features.
3. Soil and rock structure.
4. Vegetative cover.

And of these, the only one over which he has any control is the vegetative cover.

The distribution, intensity, and form of precipitation are important conditions affecting floods. Watersheds with normally heavy rainfall would seem to suffer most, but the manner of its distribution (as gentle rains or as torrential showers) is sometimes more important than the total amount. When a large percentage of the rainfall comes in the form of snow, the danger from floods is not so great.

The study and classification of many decades of rainfall data collected throughout the United States give us normal rainfall types, each of which has its definite months of maximum and minimum precipitation. Floods usually occur during or following the months of maximum rainfall, though the melting of accumulated winter snow in spring is often an important factor as well.

The Atlantic Type shows a very nearly equal distribution of rainfall throughout the year, the maximum usually occurring in July. This type dominates the Atlantic Coast, excluding New England, as far west as the upper Ohio River. The New England Type, a slight modification of this, has in addition a November maximum. Since, however, there is such an even distribution of rainfall throughout the year, floods which occur mostly in the spring must be due largely to the melting of the accumulated winter snows which are especially heavy in the mountains of this region.

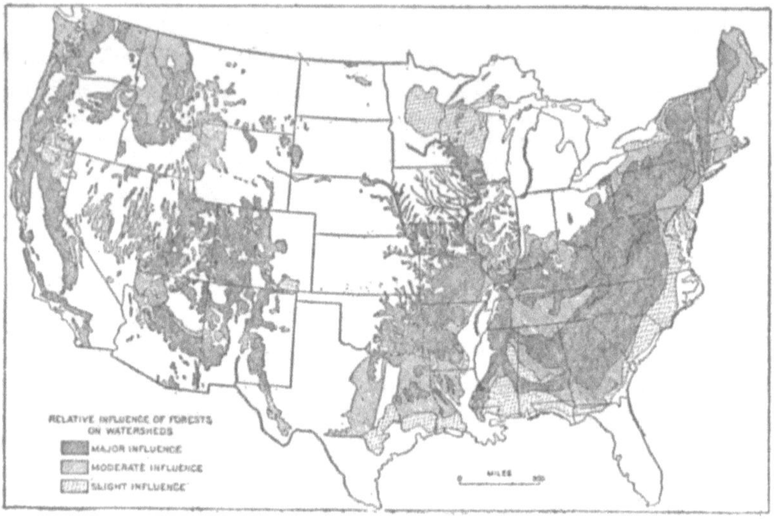

U. S. Forest Service Photo

FORESTS OF SLIGHT, MODERATE, AND HEAVY INFLUENCE ON WATERSHED PROTECTION

About three-fourths of our total forest area (450,000,000 acres) has been classified as water-protection forest, that is, as having major or moderate influence on run-off and erosion due to steepness of slope

The Missouri Type, the most important of all in point of area dominated, covers the upper half of the great Mississippi Valley from Minnesota to northern Texas and from Montana to Indiana. In this area about 70 per cent of the rainfall occurs from April to September, with the maximum generally falling in June. Floods in this area occur usually during late spring or early summer.

The North Gulf Coast Type has a summer maximum, and minima in May and October. This is modified along the South Atlantic Coast by the late summer hurricanes which delay the maximum until early autumn.

In the western plateau states, embracing the Rocky Mountains, the maximum run-off comes in early spring or early summer. This maximum run-off is greatly augmented by the melting of accumulated winter snows in the high mountains. To make matters worse, the eastern slope of the mountains is at about the same time visited by torrential downpours falling on land practically bare of forest vegetation. This helps to explain the flood waters which occur almost annually in the western tributaries of the Mississippi.

The Pacific Type, which dominates the area draining into the Pacific Ocean, is characterized by abundant winter rainfall and an almost total lack of summer precipitation. Over 90 per cent of the rainfall occurs from November to May. The dry period increases in duration towards the south. Floods usually occur during the rainy season or directly following it. Here also the spring rains are greatly augmented by the melting of accumulated snows which in the Cascades and Sierra Nevada are particularly heavy.

Heavy rainfall of short duration often causes a local flood because most of the water does not have time to soak into the ground. These are not confined to any particular region, and apparently may occur at any time during the year. Re-occurring torrential downpours occur in the upper Missouri Valley, where they often lead to high flood waters of short duration.

The form of precipitation is often important. In most of our mountain regions the precipitation is much greater than in the surrounding lowlands, and a much larger portion of the total comes in the form of snow. New England, the northern Appalachians, the northern Rocky Mountains and the Pacific mountain ranges have abundant snow. If conserved by the forest cover, the delayed snowmelt affords valuable water for summer streamflow; if unprotected, it rapidly melts and adds to spring and early summer floods. Warm rains and high temperatures combine to make a flood hazard in all mountainous regions, especially when combined with frozen ground which allows practically no absorption. Sub-freezing temperatures at night, however, are favorable for holding back the flood waters.

The physiographic factors which affect floods are, slope, aspect or exposure, and the general form and orientation of the watershed. Of course, the steeper the slope the more rapid the run-off and the less opportunity afforded for the water to soak into the ground as it runs over it. Extremely flat terrain may lose as little as 5 per cent of the average annual precipitation in the form of surface run-off, while rugged mountain terrain may lose as much as 50 per cent, other factors being equal.* Aspect or exposure affects run-off, especially in early spring when snow on a south or warm exposure thaws

* A. E. Parkins and J. R. Whitaker, *Our Natural Resources* (New York, 1936), p. 387.

out rapidly, while the northern exposure, tilted away from the warm sun, remains frozen. In New England, for example, snow conservation on these warm southerly aspects is important, and this can be accomplished only by increasing the density of the forest growth.

A wide drainage basin, covering considerable latitude north to south, may have a number of large tributaries that increase the danger of flood on the main stream, especially when the crests of the several tributaries coincide in the lower course of the main stream. Again, a river system may run parallel to the path of the cyclonic storms so that different portions of its watershed receive the rains successively as the storm area moves over them, resulting in a succession of small floods. If the watershed runs at right angles to the path of the storms a smaller portion of the watershed would be affected at any one time. The Ohio meets both unfavorable conditions of coinciding crests and unfortunate orientation.

Soil and rock conditions in a watershed also enter into the flood situation. The more water the soil will absorb, the less will run off the surface. Coarse sandy and gravelly soils permit more absorption than heavy clay soils. Forest soils with a high percentage of organic matter and a characteristic loose, porous, structure permit greater absorption than mineral soil alone. Glacial drift is favorable for delaying run-off, first on account of the numerous lakes and swamps in glaciated areas which act as storage reservoirs, and secondly because the mixed character of the soil is favorable for deep-soil percolation. Sedimentary rock is more easily weathered than igneous or metamorphic rock. The former usually gives rise to deep soils and the latter to shallow ones which, of course, lack the capacity for storing water. The erodability of the soil in the watershed may aggravate floods in that the stream channels fill with erosion debris, causing the water to spread out over the lowlands.

The vegetative cover, as already pointed out, has a very important bearing on surface run-off and hence upon floods. Data gathered by various governmental agencies, under such widely different conditions as are found in the states of Oklahoma, West Virginia, California, Idaho, Mississippi, North Carolina, Tennessee, Texas, and Missouri, have shown fundamentally uniform results on this question. Water run-off is materially decreased (other factors being the same) as we proceed from bare soil, cultivated fields, and grasslands to brushlands, cut-over forest, and well-stocked forest. The ratio of run-off from bare land, cultivated land, or forest land heavily cut over, as compared to that from well-stocked forest, ranges from 20 to 1 to as much as 100 to 1. Furthermore, U.S. Forest Service experiments in Ohio have shown that the upper three inches of hardwood forest soils may absorb fourteen times as much water as adjacent agricultural soils. The greater capacity of forest soils for absorbing protracted rains is obvious.

It is generally agreed among flood-control students that floods have their origin when the water derived from run-off and underground sources spreads beyond the restricting banks of the stream. Man cannot control precipitation, physiographic features, or soil and rock structure, but he can control the amount and kind of vegetative cover on a watershed. This, therefore, is the vital factor. All other conditions being equal, the watershed having the greatest area in forest will have the least run-off, and that same watershed will also have the smallest amount of erosional debris in the bed of the stream, and hence that stream will be less likely to spread beyond its restricting channel.

Flood Control Methods

General methods of flood control, both past and present, include:

1. Channel improvements.
2. Levee protection with spillways and floodways.
3. Retarding basins and storage reservoirs.
4. Restoring vegetation on watersheds.
5. Minor up-stream structures.
6. Improved land-use technique.

One of the first methods to be employed consisted of various channel improvements. These included the removal of brush, trees, and erosion debris from the bed of the stream to facilitate the discharge of the flood waters. This usually results in a straightening of the channel by its cutting across curves in the stream. Thus the gradient of the river and the velocity of the flow are increased, with a greater capacity for transporting debris. If this results in more scouring of the bottom and sides and the deposition of larger amounts of debris further down the river, it may be a great hazard to communities lower down the valley.

Sooner or later this method must be followed by the building of parallel dikes or levees to confine the river to its new channel. But in times of flood this too may be a hazard in that the increased velocity and carrying power of the water may scour out the levees at their bases or, by over-topping them, create rifts and crevasses which will result not only in flooding the lowlands behind the levees, but also in actually destroying the levees themselves.

This breaking of levees and dikes has led, in the lower section of the Mississippi River system, to the use of floodways to lead off flood waters behind the levees through the lower back swamp areas and away from the main stream. The old crevasses in the levees are used for spillways. In this way the flood crest of the main stream is reduced.

Although the first levees were constructed at New Orleans as early as 1717, it was not until 1879 that Congress appointed the Mississippi River Commission and authorized a levee system. This Commission established a grade line for levees from Cairo, Illinois, to the Gulf. But the levees were not high enough and the river broke through. The grade was raised five feet in 1898. The big flood of 1913 wrecked this system of levees. In 1914 the levees were again raised with what was thought to be a good margin of safety. The disastrous flood of 1927 washed over the dikes and inundated thousands of square miles of lowlands as has already been stated. Another raise of from five to ten feet was asked for in addition to floodways in back of the levees with spillways to relieve the main stream.

Up to 1932 Congress had spent over two billion dollars for river and harbor improvements, most of it for levees. The river bottom keeps on rising because of the erosion debris, and the levees are built higher and higher at ever-increasing costs. The Mississippi River now has over 1,800 miles of levees averaging twenty-one feet in height. In a number of places the river bed is actually higher than the surrounding lowland. It does not seem that levee construction has permanently solved the flood problem on the great river. The stage is all set for another "greatest flood in history."

Up to 1927 the Mississippi River Commission had stood very rigidly for levees as the only means of flood control. A number of years before, however, other methods were brought into use in state projects. Retarding basins on the tributaries of the main streams, it was thought, would check the run-off, thus flattening out the crest of the flood in the down-stream areas. Such dams are usually constructed of earth except for the gateway, and when equipped with control gates, a series of them along a stream may hold back local flood waters behind the dams which are later discharged automatically.

Such retarding basins were used in the Miami Conservancy District of Ohio, following the disastrous Dayton flood of 1913. Five such dams were constructed across the Great Miami River and its tributaries. These, together with the channel improvements and levees, have successfully stood the test of several floods. In the Muskingum Watershed Conservancy District of Ohio a similar method of flood control was used by installing storage reservoirs. This method of control by storage is expensive unless other uses like water supply, irrigation, navigation, electric power, or recreation can be coördinated with the flood control use. Such storage reservoirs not only reduce the flood crests but also permit the raising of the low water stages of the river for other uses.

A list of federal dams in process of construction or already completed, published by the New York *Times* in February, 1936, showed thirty-seven dams, with more than half of them having flood control for their purpose. Most of them, however, were multi-purpose dams.

Thus, in summing up, we may say that up to the year 1936 our national flood-control program comprised the use of levees, dikes, floodways, spillways, retarding basins, and storage reservoirs. This is known as the "down-stream" or engineering method of flood control. This policy almost completely ignored "up-stream" or vegetal control, the building of minor up-stream structures, and the correction of wrong land practices leading to erosion. It ignored the important part that forests and other forms of vegetation play in regulating stream flow and preventing erosion. Worst of all, it did nothing to alleviate local floods in up-stream tributaries which have for years caused a tremendous aggregate of damage. Lastly, it ignored the fact that destructive grazing and agricultural methods caused much of the erosion, which in turn was such an important factor in flood control, since erosion debris fills the river channels so that they cannot carry off flood waters and also eventually nullifies the effect of storage reservoirs for retarding flood-waters.

The controversy between these two schools of thought reached its climax in Europe quite a number of years ago. In recent years the pendulum has swung towards vegetal control. In these countries there is quite general acceptance of the government policy that land management—grazing regulation, forest protection, reforestation, agricultural practices which preserve the soil, and the building of minor up-stream structures—must be an essential part of a complete water-control program. This must be coördinated with the down-stream structures already described.

France, Switzerland, and Italy have been able to attack the problem from this double point of view and have been able to minimize flood damage by coördinating the two programs. Engineers, foresters, agronomists, farmers, livestock producers, and lumbermen up to the war coöperated to get the maximum returns from the investment of every flood-control dollar. That is exactly the policy which must be adopted in our own country if we are to bring order out of chaos.

Recent Flood-Control Legislation *

As late as 1927 the Mississippi River Commission rejected reforestation as a method of flood control in the lower Mississippi Valley. But the following year the Mississippi Flood Control Act authorized the Secretary of Agriculture to study ways in which forestry might aid in the control of Mississippi floods. The same year the McSweeney-McNary Act authorized investigations to determine methods of preventing erosion and controlling water flow. In 1929 Congress provided field research stations to develop measures to prevent erosion and retard run-off. In 1933 relief legislation authorized work to prevent soil erosion and floods, and in the same year the Soil Erosion Service was established from the small beginning made in the Bureau of Chemistry and Soils several years before. With the establishment of the Soil Conservation Service in 1935, Congress declared it to be the policy of the government to provide permanently for the control and prevention of soil erosion and the consequent control of floods and preservation of our soil resources.

All these acts were only a beginning, but the "up-stream" advocates had an entering wedge. Valuable data on erosion and floods, and especially reconnaissance surveys upon the character and extent of erosion throughout the United States, were the direct result of this early legislation.

Before we look into the watershed measures to be adopted under the current Flood Control Act, let us take a fleeting glance at the fundamental reasons behind these measures. Let us get some idea of the actual condition of the surface area of this country. If we are looking for reliable data to show the need of a broader and more progressive government land-management policy, we can find plenty of it in recent government reconnaissance surveys:

50 million acres of abandoned crop land have lost all their top soil by erosion.

100 million acres of crop land still under cultivation have lost all or a greater part of their top soil.

One-half of the nation's better crop land is affected by erosion.

100 million acres of cut-over and burned-over forest land is not restocking and is contributing very appreciably to the flood and erosion problem.

850 thousand acres of forest land are damaged or devastated every year under the cut-out and get-out system.

* *The Land in Flood Control,* U. S. Department of Agriculture, Misc. Pub. 331, 1938.

40 million acres of forest land are swept by fire annually destroying all or part of the protective vegetation.

580 million acres of western range lands show more or less serious accelerated erosion and evidences of more rapid run-off.

In terms of feed for sheep and cattle, 53 per cent of our western range lands have had their carrying capacity reduced anywhere from 50 to 100 per cent.

Adding all these items together, does it make the picture any clearer by stating that this destruction endangers almost 50 per cent of the area of the continental United States? Could any flood control policy conceived by a great national government ignore these incisive facts?

The Flood Control Act of 1936 provided that "federal investigations and improvements of rivers and other waterways for flood control and allied purposes shall be under the jurisdiction of and shall be prosecuted by the War Department" and that "federal investigations of watersheds and measures for run-off and waterflow retardation and soil erosion prevention on watersheds shall be under the jurisdiction of and shall be prosecuted by the Department of Agriculture." In the Act there was designated a long list of localities in which the Secretaries of both Departments are to conduct preliminary examinations and surveys, and $10,000,000 was appropriated for this survey work.

In the following year this was amended so that the work of the Agricultural Department would be confined to those watersheds and rivers on which the War Department had been authorized to work. Appropriations for carrying on examinations and for prosecuting flood-control measures were extended to June, 1944. In order to secure further coöperation and coördination of effort, the plans approved by the Secretary of Agriculture are submitted to the President, who designates the order of urgency of the projects and correlates them with the work of the War Department.

Each watershed survey must furnish exact information as to the social, economic, physical, and climatic factors involved. A complete study is made of the soils, especially their texture, depth, erodability, infiltration capacity, and productivity. Great importance is attached to present land uses and topography, as to whether the crops are soil-protecting crops or soil-depleting crops, and the historical development of land uses. A complete history of erosion and floods on the given watershed is worked out. The surveys will reveal the extent, location, density, composition, and age class of the forest cover, its fire history, and all conditions affecting erosion and stream flow. These surveys reveal the tributaries which are contributing the most run-off and the most silt, and these critical areas are given first consideration. Rarely will it be necessary to treat a whole watershed with an equal degree of intensity.

Coördinated Land-Use Technique *

All who use land or soil must coöperate in better land-use technique. Land use, erosion, and flood control are inseparable. Flood-control measures on land divide themselves into four categories: (1) for agricultural lands, (2) for forest lands, (3) for range lands, and (4) in streams.

For soil erosion and water-control on agricultural lands, the Department of Agriculture has access to the experimental data of the Soil Conservation Service on its more than 175 demonstration watersheds. Prominent among the techniques that have been worked out are contour farming, terracing, strip cropping, crop rotation, and pasture furrowing. Such measures are designed to hold back water flow and thereby permit a more complete infiltration of water. In contour farming † each furrow follows a horizontal path so that it becomes a small dam or terrace rather than a channel for the run-off of water, as would be the case if the furrows ran up and down hill. Strip cropping makes use of the fact that close-growing vegetation retards water run-off while row crops increase it. Two kinds of crops are alternated in strips, thus close-growing crops like alfalfa and clover retard the run-off from row crops, allowing more water to be absorbed. Terracing of farm land permits more rapid infiltration of water into the soil and prevents the quick concentration of water into narrow channels which leads to the formation of gullies.

Since from one-fourth to one-third of the Continental United States will probably always remain absolute forest land, the control of erosion and floods on this type of land is of the utmost importance. The forester believes that control work on forest land is immensely important because most of our absolute forest lands are on steep slopes which favor rapid run-off, because most of these forest lands are at the headwaters of our large rivers where floods are born, and because research has shown that the forest cover is the

* *The Land in Flood Control,* U.S. Department of Agriculture, Misc. Pub. 331, 1938.

† Farmers in the Piedmont Region of the southeastern states have followed a rudimentary form of contour farming and terracing for at least a hundred years.

most effective cover to retard run-off and stop erosion. In the management of its National Forests, the federal government has developed over the last thirty years many water-control techniques. Protection from fire is, of course, paramount. Reforestation by planting and seeding is used to restore a forest cover on denuded land and to start forests where the land is inadequately protected. Measures to encourage immediate natural reproduction are used wherever timber sales are made. But control work may involve the planting of shrubs and grasses as well as trees.

Where organic topsoil has been lost, grasses and shrubs may be used to prepare the way for trees. In some badly burned forest areas, the sowing of a cover crop has been used to stabilize the soil for a brief time until natural vegetation can get started. Over 150 plants have been tested by the Forest Service. In California, black mustard seems to give the maximum of protection in the shortest time. Following a devastating fire in the National Forests of southern California, this seed was sown from airplanes or by hand.

Mechanical barriers to water-flow on forest land also have been used. Contour terraces and trenches have been found to be useful on critical, flood-producing forest lands in the West.

Flood and erosion control methods on range lands are important on account of the vastness of the area affected—almost one-third of our total land area. Overgrazing is the primary cause of grassland depletion. The native perennial grasses and browse plants that once maintained the soil in good physical condition to absorb water have almost disappeared from large areas of western range lands. Superior soil-holding species have been replaced by less valuable kinds. Erosion has not only removed the rich, surface layers of soil but has also filled near-by irrigation ditches and reservoirs with silt. The western live-stock industry and western irrigation agriculture are absolutely dependent upon this forage and water. These native plants may be brought back to the depleted lands by improving the range-management practices. Lowering the quota of stock allowed to graze on an area may give the grasses a chance to recuperate. On highly critical areas all stock must be excluded. On some of these the more expensive methods of control may have to be used, such as contour furrowing, range terracing, building check dams across gullies, and the revegetation of the land by seeding of forage grasses.

Flood-control measures in the streams must supplement those on the watersheds. The Agricultural Department is using up-stream engineering devices and working from the headwaters towards the large, down-stream structures of the War Department.

Up-stream engineering structures are simple and relatively inexpensive, consisting of check dams, ponding basins, and desilting basins. In the upper portions of watersheds the purpose of these structures is to arrest the destructive force of the water. Farther down-stream protective structures will be installed to stabilize stream banks, stream beds, and flood plains. Whenever possible vegetation will be used in this stabilization work.

The final stage of the federal flood control program is coöperation with state and local agencies. After the federal government has decided that federal assumption of the cost and responsibility of control is warranted by the conditions and the benefits to be derived, it is necessary to find out whether state and local agencies would be willing to coöperate. Much of the control work will have to be done on private lands and much of it must be followed by improved practices in agriculture, forestry, and grazing on the part of the owner of the land. Furthermore, land-use practices and conservation measures are of such a nature that they call for constant attention and inspection on the part of the individual using the land. In other words, aside from spending government funds, state and local authorities must guarantee that the watershed projects completed will be properly maintained.

To achieve this coöperation, the Soil Conservation Act of 1935 authorized the Secretary of Agriculture to require the passage of state laws for the prevention of soil erosion as a condition to the extension of benefits under the Act to lands in any state. A model state soil conservation districts law was drafted by the Department of Agriculture, and by July, 1938, about twenty-six states had passed laws modeled after this standard law. Similarly, in the 1937 amendments to the Flood Control Act of 1936, Congress authorized the Secretary of Agriculture to require the passage of legislation dealing with flood control by the separate states, as a condition to securing federal aid. At the present time the policy of the Department with regard to lands that are in private or in non-federal ownership is that preference will be given to those states that have passed flood-control legislation, and within the states preference will be given to areas located in soil conservation districts.

The soil conservation districts laws provide for a local organization of land users for the purpose of adopting regulations which seem to be necessary for their own interests and those of the general public. The abuse of land by improper lumbering, agricultural, and grazing methods can be stopped if a sufficient number of land owners of a locality will form an organization under existing laws. The districts have the power to carry out erosion and flood control measures.

State zoning laws provide a method by which local people may exercise restrictions over the use of land, much as city zoning ordinances do in our urban centers. Such laws may reserve certain areas for forestry, recreation, and watershed protection and may prevent the further use of lowlands subject to severe flood damage.

Summary

From the preceding discussion, certain inescapable facts stand out. For over a hundred years certain European nations have based their flood-control policy on the protective influence of forests. It has been recognized that these do double duty in decreasing run-off and erosion and in keeping the stream channels clear of erosion debris. In America we were slow to accept these facts; but by the establishment of federal forest reserves during the last fifty years, by the setting aside of many state forests, and by the enactment of basic forest laws, we have at last recognized these fundamental forest-water relations.

The record shows that this "up-stream" or "vegetal control" method has, until recently, been ignored by our flood-control engineers and commissions. They steadfastly adhered to the use of levees, dikes, and floodways, or what is known as the "down-stream" method of control. Levees were built higher and higher in the face of ever-increasing flood disasters. In 1937 the Ohio equalled previous estimates of the maximum flood considered possible. And the end is not in sight.

The turning point in our philosophy of flood control came when the two schools of thought were combined in the Flood Control Act of 1936. Henceforth, it is the declared policy of the federal government that down-stream control by the use of levees, dikes, floodways, spillways, retarding reservoirs, and storage basins is to be coördinated with up-stream measures consisting of forest protection, reforestation, grazing regulation, restoration of natural plant cover, better agricultural practices, and minor up-stream engineering structures. Destructive land-management policies of the past have contributed very largely to land erosion and floods, and it is vitally necessary that these "horse and buggy" methods be replaced by a progressive and coördinated land-use program, looking ahead at least 100 years.

Such a program must necessarily focus a large part of its attention on the absolute forest land because of its enormous area, its steep, erodable slopes, its location at the headwaters of all our large rivers, and because on such lands the forest cover is the most efficient cover for preventing erosion and floods.

Finally, foresters do not claim that proper care and use of forests will prevent all floods.* Certain combinations of soil and weather conditions will produce a flood despite the beneficial action of the forest cover. It must be noted, however, that flood history has shown that the difference between the high-water mark of a flood that becomes a catastrophe and one that does little or no harm may be but one foot or less. Many streams rise to the top of their banks and do no harm. It is the surplus that goes over the banks that does the damage. It is by retaining this often harmful surplus that the forest plays its greatest rôle. The great reservoir capacity of the forest soil and forest floor has in many cases proved to be the all-important safety factor in preventing flood disasters.

Several thousand years ago an old Chinese sage warned that "the mightiest rivers are cradled in the needles of the pine trees." That is still sound philosophy.

* *Forest Conservation: A Guide to the Social Studies* (U.S. Forest Service, 1937), p. 70.

CHAPTER VIII

LUMBER AND ITS BY-PRODUCTS

WOOD IS THE MOST adaptable material known for the various arts and industries. Weight for weight, it is stronger than metal, more easily workable, and lends itself to manufacture in any sort of shape or design. Still comparatively cheap as a construction material, it is available in many kinds, forms, and sizes. The wood of post oak is so hard and heavy that it will not float in water—white pine and yellow poplar are light but strong. Hickory is very tough, but will not stand weathering—so it is good for tool handles, poor for fence posts. Cypress wood is so resistant to decay that its bored-out logs have been laid underground for water mains. Black walnut and holly may be used for the contrasting squares of checker boards—one very dark, the other satiny-white. In its infinite variety of grain and texture, wood is invaluable for the decorative and utilitarian arts. No other material is capable of taking such a beautiful finish.

More than four thousand uses of wood are recognized, and others are being constantly developed. There are wooden mine shafts and wooden airplane parts; wooden pavements and floors, and linoleum made from sawdust; wooden wheelbarrows and freight cars, toothpicks, tobacco pipes, artificial legs, and telephone poles; wooden barrels, boxes, crates, and hampers. Sausage casings are made from wood, and so are fine hosiery and expensive tapestries. Rayon, cellophane, doorknobs, artificial leather, dress goods, radio cabinets, and thousands of other products are now made from wood in the form of cellulose plastics. In short, modern civilization, in spite of the many substitutes that have been devised, is still surprisingly dependent on wood and its by-products.

Our total annual wood consumption is almost beyond mathematical comprehension. The amount used would probably build a double row of contiguous seven-room houses around the earth at the equator. Government figures tell us that in the period from 1925 to 1929 it was over sixteen billion cubic feet a year. This figure includes losses from fire, insects, and diseases. We are consuming wood 1.2 times as fast as it is growing and we are consuming saw timber 1.5 times as fast as it is being replaced. Annual growth over the country is estimated at about eleven billion cubic feet per year.

Let us look at some more figures. Of the 822,000,000 acres of original forest, only about 100,000,000 acres remain stocked with old growth. In other words, the nation has spent 88 per cent of its wood capital on the basis of area cut over. Counting what second-growth has come in, there are still about 519,000,000,000 cubic feet of timber remaining—about one-third as much as the original supply—but much of this is inferior in quality to the original stands, and is located far from the centers of consumption.

How long will the remaining supply last? That, of course, depends upon the future consumption of wood products, losses from fire, insects, and diseases, and the demands of the export markets—all of which are impossible to predict accurately. Careful estimates made by the government place the annual future drain for some time to come at between fifteen and twenty billion cubic feet. While our lumber consumption dropped sharply between 1930 and 1940, it has risen rapidly during the war and there is every reason to believe that building operations after the war will rise from two to three times the pre-war level. Numerous building experts connected with corporations manufacturing building materials have estimated that from 1,600,000 to 2,000,000 new dwellings will be needed each year for at least ten years after the war is over. The export demand for lumber to rebuild Europe will be tremendous. All of which seems to indicate that the drain on our forests will continue.

If the growth-drain figures above are correct, it would indicate the end of our accessible virgin supplies in another generation. Since it takes from 50 to 120 years to grow saw timber, it does not seem that we can expect much help from our second-growth forests until the end of the present century. There need be no timber famine, since annual growth will tide us over, if we do not deplete our second-growth stands. There will, however, be a shortage of certain valuable species, especially certain hardwoods, and certain higher grades and large sizes of saw timber.

The lumber industry, one of our greatest and most important, has had a conspicuous part in building American civilization. Over ten billion dollars are invested in forests, machinery, sawmills, and wood-working industries. Next to food, shelter is the most essential human need. After agriculture, the forest is our greatest single resource, its products being valued at three billion dollars annually.

This industry, also, has always been a great employer of labor. The management of forest lands and the manufacture and distribution of forest products in the entire United States were estimated to supply, directly or indirectly, the equivalent of year-long employment to 3,800,000 people in 1936.* More than one-tenth of the country's population is dependent more or less for support upon the forest and its numerous products.

A look back into our history reveals lumbering as our pioneer industry, even preceding agriculture. It had in it the spirit of adventure and exploration, for it helped to tame the wilderness and it cleared the land for agricultural settlement. The western march of settlers was necessarily predicated on the exploitation of the forests, for transportation and communications systems, as well as cities, towns, and farmhouses, were dependent on the use of wood.

More than this, the lumber industry personifies the triumph of mechanical skill and ingenuity over the mighty forces of nature. For over a hundred years it has had at its command unlimited energy and resourcefulness—the very qualities which our forest conservation movement needs at the present time. It developed logging machinery unique in forest industry anywhere in the world. Forests were removed from every conceivable kind of terrain; no mountain was too steep, no valley too deep, and no swamp too wet. Logs travelled over the ground, through the water, over snow and ice, and through the air. Logs weighing tons were handled like match sticks. Over a period of three hundred years the old sash-saw mill run by water power, with its output of a few hundred feet a day, has gradually evolved into the gigantic band-saw mills with shotgun steam feed, capable of turning out a million feet a day.

The history of the lumber industry in America reveals the shifting areas of operation. When the colonists landed early in the seventeenth century, they were confronted with an uninterrupted wilderness of forest. Clearing began in order to make room for agricultural pursuits. Huge trees, which would now be worth a fortune, were felled in piles and burned. The first sawmills appeared in Jamestown, Virginia, in 1625 and in Maine in 1631. These were of the single sash-saw type driven by water power. Before this time timbers were hewn, shingles were split, and boards were whip-sawed. All the work was done by hand with axe, wedge, saw, and similar tools. Ship spars, ship timbers, staves, and lumber were produced for local needs, and these products also found their way into the European markets. The few sawmills existing up to the beginning of the nine-

* *Forest Lands of the United States,* House Report 323, 1941, p. 18.

teenth century did not make much impression on the enormous stocks of virgin timber. It was not until the use of steam in the sawmills that cutting really began.

Maine was the greatest lumber-producing state in the Union until 1850. In that year New York supplanted Maine, and the northeastern states produced almost 55 per cent of the lumber output. In 1860 Pennsylvania ranked first, white pine being the principal species cut. By 1870 the shift was to the great white-pine belt of Michigan, then probably the greatest continuous belt of high grade pine in the world. The peak of production in the Lake states was reached at the end of the decade of 1880-1890 and has declined ever since. The next shift was to the great southern pinery. Since 1895 this region has produced from 20 to 40 per cent of the total annual cut. The peak was reached here in 1919, and then slowly began to decline. By 1929 the western states produced over 43 per cent of the lumber and have been in first place ever since. In that same year the West and the South together produced over 85 per cent of the country's total cut. That the western states will remain the principal source of our lumber supply is evident from the fact that at the present time over 60 per cent of the country's virgin saw timber is in the Pacific Coast states.

The nomadic character of the industry brought with it a floating population that moved with the sawmills from place to place. Large towns would come and go. Virgin forests would give way inside of a few weeks to a rough-and-ready lumber town, and as soon as the region had been exploited, empty shacks and piles of sawdust would mark the place where 15,000 to 20,000 people had recently lived. The smaller units of government found that not only was their taxable property gone but the taxpayers had vanished also. The thousands of acres of cut-over and burned-over land were often taken by the state for unpaid taxes, and the county was left without a source of income to run the schools and build the roads. In other words, all the permanent good things of a civilized community—schools, churches, libraries, substantial homes, and settled industries—were not a part of this old system of "cut out and get out." This was the worst feature of this great industry, a sharp contrast to the healthy, permanent lumber industry which was developed in Europe and which must eventually be developed in this country if we expect it to survive.

During the last twenty years, as the more valuable species grew scarcer, the less valuable ones began receiving attention. As oak, white pine, spruce, and yellow pine became depleted in their respective localities, the so-called "inferior" species such as hemlock, red gum, lodgepole pine, incense cedar, California white fir, and many others came into the market, many of them under "assumed" names.

Keen competition has compelled great advances in the handling, manufacture, and distribution of lumber. Great strides have been made in grading lumber, in the establishing of national organizations of lumber manufacturers, and in advertising the merits of different grades and species. There has developed more coöperation among the producers, and also between the lumber industry and the Forest Service.

Most important of all, however, in the last two decades there has been a shift of sentiment in favor of forestry, due to the realization that the end of the virgin supply is in sight. The Forest Service has greatly assisted the industry by its research in the physical and mechanical properties of wood, and in the seasoning and preservation of lumber and timber. This work was begun by the Forest Products Laboratory in 1910 and has done more than anything else to bring the Forest Service and the lumber industry into closer understanding. Further evidence that these two groups are working together can be found in the fact that numerous timber owners, associations, private forest protective organizations, and wholesale and retail lumber manufacturers' associations now employ foresters in high executive capacity.

But the last forest frontier has been crossed. The last great reservoir of raw materials has been tapped. The handwriting is on the wall! The industry's nomadic existence is at an end. There is no place to begin new operations. Our ingenuity and enterprise must now be turned in a different direction. A great period of adjustment is at hand—not only physical and mental adjustment, but spiritual adjustment as well. Present profits must in part be sacrificed to future security. The old traditional attitude of regarding the industry as a "mining" enterprise must be abandoned. The growing of timber must from now on precede the cutting of it—if there is to be any to cut.

The Use of Lumber and Other Wood Products

The average annual production of lumber in the pre-depression years 1925-1929 was about thirty-eight billion board feet; in 1936 it had dropped to under twenty-eight billion board feet. During the war years, production has gone up again, almost reaching the pre-depression level.

Usually about two-thirds of our lumber goes into construction, mostly residences and other forms of building. We must remember that four-fifths of our dwellings are of all-wood construction and that they house over ninety million people. Prospective needs are great, for it is reported that one-third of our citizens live in substandard houses and over 60 per cent of our dwellings are in need of repair. The rest of the lumber goes into such products as furniture, coffins, refrigerators, novelties, and woodenware—not to speak of ships, airplanes, and weapons of war. Much of our lumber would be wasted were it not for the manufacture of articles from odd-sized pieces and from the cuttings and trimmings left over from the making of larger structures and commodities. There has been much progress in developing methods of making wood resistant to decay, insect attack, shrinkage, swelling, and fire. These processes lead to more efficient utilization, and must be classed as true conservation measures.

Besides wood cut for lumber, nearly as great a volume is used in other forms, as shown in the following table of timber removed from commercial forests of the United States:

Table 5

TIMBER CUT FOR COMMODITIES, 1936 *

USE	THOUSANDS OF CUBIC FEET	PER CENT OF TOTAL
Lumber	5,367,585	47
Fuelwood	3,619,482	32
Pulpwood	705,924	6
Hewed ties	354,189	3
Fence posts	327,060	3
Veneer logs	252,443	2
Mine timbers (round)	161,016	2
Cooperage stock	149,447	1
Shingles	108,658	1
Other	354,203	3
TOTAL	11,400,007	100

* *Forest Lands of the United States,* House Report 323, 1941, p. 41.

The American people have always used wood freely. We still use one-half the lumber, one-half the paper (mostly made from wood), and two-fifths of the wood in all forms consumed in the entire world. Our per capita consumption in 1936 was over 100 cubic feet, as compared to about 20 feet for Great Britain—and next to Sweden and Finland, we were the greatest users of wood per capita in the pre-war era.

The question might naturally be raised: Will not the use of wood continue to decrease as substitutes replace it? There has, indeed, been a downward trend of wood consumption all over the world in the last quarter of a century—but this has no doubt been due, at least in part, to unsettled world conditions. Some countries, also, like the United States, have approached the limit of their agricultural and industrial expansion, by which the need of wood has been cut down. Some inroads have been made by substitutes, but these have never greatly disturbed wood consumption, because any decrease on this account is quickly offset by an increase in other uses due to new discoveries and new technological developments. A decrease in lumber consumption in the United States has been partly offset by the increased use of plywood, and of pulpwood for the fiber and cellulose industries. Indications point to a greatly increased demand for wood in the future.

Next to lumber, fuel is the most important forest product. The average fuelwood consumption has been about sixty million cords a year. This use of wood has been decreasing for the last few decades because other fuels like gas, oil, coal, and electricity have become more available. People who live in timbered regions at a distance from these fuels still rely on wood, whose use is largely local on account of its bulk. Often fuelwood is a by-product of other forest operations. As far as possible it should consist of inferior timber, or dead insect-infested trees that would otherwise go to waste. Unused tops, branches, broken timber and other logging debris should be utilized for this purpose. A great many species of wood are used. In the Northeast, hardwoods like birch, beech, oak, and maple are preferred.

Although thousands of attempts have been made, a suitable substitute for wooden railroad ties has not been invented. From 65,000,000 to 70,000,000 ties, mostly hewn by hand, are used on the 420,000 miles of our railroads. The outstanding virtues of wood for this use are its durability (when treated with creosote or zinc chloride), its resilience, or ability to resist mechanical impact, its availability, and its low cost. Oak, southern pine, Douglas fir, cedar, chestnut, and cypress are among the species principally used.

Iron and concrete posts have been used as substitutes, but as yet their use is negligible. The principal qualifications for a good fence post are durability, lightness in weight, straightness, and ability to hold a nail well. The species of wood used depends upon locality, but usually cedar, redwood, oak, locust, chestnut, cypress, and longleaf pine are preferred. Preservative treatment often doubles the life of a post. While this industry is often carried out on farmers' woodlots, it is also carried on in connection with pole production.

Pulpwood is used in ever-increasing quantities for paper, rayon, plastics, and other cellulose products. In 1938 over fifteen million cords of pulpwood was consumed, of which 46 per cent was imported.*

Formerly most of our paper was made from rags; today 85 per cent of it is made from wood. We are the greatest users of paper in the world, our per capita consumption each year being about 200 pounds. Paper-making is one of our great industries, employing about 200,000 people in normal times. Paper products are, of course, too numerous to list, but they fall into several large categories: newsprint; cartons, packing boxes, and fiber board; wrapping paper; book paper; stationery and other high grade paper; and miscellaneous products such as wall paper and fabricated articles.

Two general methods of papermaking are employed: the mechanical and the chemical. The mechanical process grinds up the logs into pulp for cheaper grades of paper, like newsprint; while the chemical process breaks down the wood with chemicals, for making higher grades of paper. The three chemical processes take their names from the chemical used: soda, sulfite, and sulfate processes. Each is adapted for using certain raw materials and for making certain grades of paper.

A new process for converting southern pines into woodpulp for newsprint, invented by the late Dr. Herty of Georgia, has been put into production. A number of other mills making kraft paper have been established in the South where the growth of pine is rapid, labor and power are cheap, water is abundant, and facilities for exporting are good. Since a paper mill is a permanent investment of from five to fifteen million dollars, and since it must get its supply of wood within a two-hundred-mile radius, sustained-yield management will necessarily have to be introduced to give the mills a continuous supply.

In the West, pulp is made largely from sawmill waste as a byproduct of the lumber industry. In the Lake and northeastern states our best pulp forests are fir, spruce, hemlock, and poplar. Cottonwood, pines, and some hardwoods are also used. Many of these big pulp and paper companies are operating upon a continuous produc-

* *Forest Lands of the United States,* House Report 323, 1941, p. 23.

tion basis and are employing foresters to supervise their operations.

In the use of wood pulp for plastics we have one of our most rapidly expanding industries. It is a field of enormous potentialities and one that conceivably may revolutionize many other industries. The numerous synthetic substances created out of natural materials may in time seriously compete with such staple articles as metals, silk, cotton, and wool—for plastics have come to be used in an endless array of tools, commodities, and gadgets. There are artificial cottons, wools, linens, and silk in great variety, timers and distributers on automobiles, radio cabinets, buttons, knobs, handles, pipe stems, beads, billiard balls, insulators, artificial leather, movie films, collodion, explosives, celluloid products, lacquers, cellophane, fountain pens, auto trimmings, fishing lines, tennis racket strings, toothbrush bristles, nylon stockings, woven glass filter cloths, and rayon cord for auto tires.

These synthetic substances can be roughly classified into three groups:

1. Synthetic resins derived from acid or mineral bases which produce such organic textiles as nylon.

2. Proteins derived from casein and soybeans for the production of artificial wool.

3. Vegetable fibers derived from wood pulp, cotton linters, corn stalks, sugarcane residue, etc., for the production of rayon filament and spun rayon products.

From the standpoint of forestry, only the last of these is important. As the other two classes of products may eventually compete with rayon, however, it will be well to look at them briefly.

Nylon is the raw material for the first organic textile fiber of mineral origin. It is stronger and more elastic than silk, wool, linen, rayon or cotton, and is said to be made of coal, water, and air. It has already become an important competitor of cotton, rayon, and silk, and before the outbreak of World War II it was used increasingly for high-priced hosiery. But besides its important uses as a textile fiber, this material can be pressed into plastics, rolled into sheets, and produced as filaments, bars, or rods. On account of its strength, durability, resiliency, and high resistance to water it will probably have endless uses. It can be made into lace, knit goods, woven dress goods, upholstery material, linings, and possibly even rugs and carpets. In time it no doubt will compete with filament and yarn rayons made from cellulose. It was first put on the market in 1940.

Artificial wool is made from such protein substances as casein and soybeans. It is said that casein wool shrinks less than natural wool but that it does not wear so well. The cost is about the price of natural wool. The Italians are said to have an artificial wool called Lanital. A product similar to this has been developed by the U.S. Bureau of Dairy Industry. Soybean wool is still experimental in both Europe and America. Germany and Japan have been developing wool substitutes from hoof and hide glues and fish glue. The product and the market are both uncertain in this plastic, but further experimentation might eventually lead to competition with natural wool, one of the main products of our livestock industry. We consumed over 280,000,000 pounds of natural wool in 1938.

About 67 per cent of all rayon produced in the United States is made from wood pulp, the rest from cotton linters. Many types of raw materials have been tried, but none has been able to compete with wood pulp, generally regarded as our most highly concentrated form of cellulose. We used 330,000,000 pounds of rayon in 1938.

The viscose process of making rayon starts with bleached wood pulp or cotton linters. This is dissolved to produce a viscous liquid and then hardened into filaments. These may be spun into thread and the thread woven into cloth. As such it is called rayon or artificial silk. In the viscose process, cellulose treated with sodium hydroxide and then with carbon disulfide is forced through tiny holes into a precipitating bath of sulfuric acid and sodium sulfate where it coagulates into solid filaments. These may be used in that form or may be spun into thread. In 1929 about 86 per cent of the world's output of rayon was made by this process. If the solution is run through a slit instead of fine pinholes, it hardens into sheets and forms cellophane. The process of making rayon of the continuous filament variety is about fifty years old; only within the last six or eight years has yarn or spun rayon come into use.

The cuprammonium process uses wood pulp or cotton linters mixed with copper hydroxide, and the mixture is dissolved in concentrated ammonia solution. After cleaning, the liquid is forced through spinnerets into an acid or alkali bath and the cellulose is precipitated as a thread.

One of the most important of the new synthetic textile materials is rayon staple fiber, or spun rayon. Most of the viscose variety is made from wood pulp. Some acetate rayon, made by combining cellulose with acetic acid, is made from cotton linters, but the acetate rayon is more expensive than the viscose variety. Spun rayon is made to imitate woolens, linens, and quality cottons and can be spun with an admixture of each of these. In general, it is cheaper than the filament variety. It is a great competitor of silk in women's dresses, but does not compete with cotton, as its price is too high. The use of rayon staple is increasing rapidly.

Besides being used in textiles, cellulose plastics made by the viscose, cuprammonium, and acetate processes have a wide variety of other uses, as already indicated. Then there are also the so-called nitro-celluloses, made by the action of nitric acid on pure cellulose. This process produces raw materials for smokeless powders, pyroxylin, celluloid, explosives, moving picture films, artificial leather, and nitro-cellulose lacquers for automobiles and furniture (Duco).

An enormous amount of wood goes into the construction of barrels, buckets, kegs, and similar containers. Barrels usually have three distinct parts: the staves, the heading, and the hoops.

Tight cooperage staves are used for barrels, casks, and other containers holding liquids. They are cut from bolts of high-grade timber, mostly of oak, gum, or ash. Bolts are also used for making the heads of tight barrels. The hoops are usually made of steel or iron.

Slack cooperage is less exacting in the quality of timber demanded, for such barrels are used to store dry materials such as sugar, flour, salt, crackers, apples, and other agricultural products. Almost any species of wood can be used for slack cooperage. Red gum, pine, beech, chestnut, basswood, and others are used for staves and headings—and elm, hickory, and ash are among the common hoop woods used.

About $45,000,000 a year is spent for timber used as supports to prevent caving in mines. During 1936 about 160,000,000 cubic feet of round mine timber were used, in addition to much sawed timber. The materials come in the form of prop timber, collar timber, mine rails, mine ties, caps, lagging, and shoring. A large part of this material is used in coal mines, but a considerable quantity is needed for mining iron, copper, and the precious metals. Many species of wood are used, of which pine, fir, hemlock, oak, chestnut, beech, maple, and other hardwoods are the most common.

About 250,000,000 cubic feet of logs and bolts are used annually for veneering. By this process wood is sliced into thin sheets, varying usually from one two-hundredth to one-fourth inch in thickness. With the depletion of our valuable cabinet and furniture woods of beautiful grain, furniture is now made with a face of veneer and a center or core of cheaper and low-grade woods. Thus our highly decorative and figured woods like walnut, mahogany, and quartered oak are more sparingly used, and the supply can be made to last longer.

Originally veneer was used principally for fine furniture and cabinet work, but at present cheaper veneers are finding new uses constantly. Thus veneers are cut not only for table tops, bedsteads, dressers, pianos, cabinets, and musical instruments, but also for doors and door panels, packing boxes, cigar boxes, berry and fruit boxes and baskets, chair seats, backs of mirrors, drawer bottoms, egg cases, wooden dishes, toys, trunks, and a great variety of novelties. Besides walnut, mahogany, and oak, the species most used are red gum, yellow pine, hard maple, cottonwood, yellow poplar, basswood, birch, elm, and Douglas fir. Almost any wood can be made into veneer if it is free of defects and meets the requirements for which the product is to be used.

Veneers are cut by three general methods: rotary cut, sliced, and sawed. In the rotary cut veneer, the log, after having been boiled for many hours, is rotated against a huge stationary knife and the effect is much like unrolling a huge roll of paper. Walnut and mahogany are cut this way. When veneer is sliced, the wood (in the form of bolts) is moved downward against a stationary knife. This method is used in making quartered oak, sycamore, and red gum veneers. When veneer is sawed, it is cut like lumber, but in very thin sheets. The highest grades of finish and cabinet veneers are made by this process.

Shingles are also sawed from bolts. Cedar, cypress, yellow pine, redwood, and spruce are used. Wood shingles demand durability, light weight, nail-holding capacity, and freedom from splitting and warping. Competition of other materials has been growing steadily. Asphalt, asbestos, slate, tile, metal, and tar are some of the substitutes used in response to the demand for fire-proof construction, especially in the cities.

Poles are used for telegraph, telephone, and power transmission lines, and piles are used for bridge, trestle, wharf, dock, and harbor construction. Of the 90,000,000 miles of telephone and telegraph wires above ground, one-half are on wooden poles. Such wood must be durable in contact with the soil, strong, light in weight, straight, and reasonable in price. Various species of cedar, chestnut, oak, pine, cypress, redwood, and Douglas fir meet these requirements. Both poles and piles are treated with preservatives to lengthen their life in service.

Since the outbreak of the war distillation wood has become increasingly important. When wood is heated in the absence of air, distilled products and residues are produced which have a considerable value in various industries. Oak, beech, birch, and maple are the species usually used. A cord of hardwood will yield 10 gallons of wood alcohol, 25 gallons of tar and grease, 200 pounds of acetate of lime, and 1,000 pounds of charcoal. The crude alcohol is refined into pure wood alcohol, acetic acid, and acetone. The distillation of pine wood yields turpentine, pine oils, rosin, tar, and charcoal, all used in various industries. All these products have many important war uses, and recent research has added to the number.

The species most used for excelsior are cottonwood, aspen, yellow pine, basswood, yellow poplar, and white pine. Wood for this purpose must be resilient, odorless, straight-grained, light in color, and light in weight. Excelsior is used in the upholstery, mattress, automobile, and furniture industries, and for packing breakable goods such as china, glassware, and toys. Wood wool, a very fine grade of excelsior, is used for filtering purposes.

Maple Syrup and Sugar

This industry is small but important in the northeast and Lake states, the top rank in production being held by Vermont, New York, Pennsylvania, Ohio, Michigan, and New Hampshire. It is carried on largely by private farmers on their own land, and in 1936 about 1,000,000 pounds of sugar and 500,000 gallons of syrup were produced, having a value of $4,000,000.

All our thirteen species or varieties of native maples produce a sap that can be reduced to a syrup or sugar, but only one species, the hard or sugar maple, is of commercial importance. It produces 80 to 90 per cent of the total product. The season varies from about March 15 to April 15, depending on the region and the severity of the previous winter. The average tapping period usually varies from thirty to forty days. Individual trees produce from one to seven pounds of sugar per season. About thirty-two gallons of sap are needed for one gallon of syrup, and from four to five gallons of sap for a pound of sugar.

Heavy sap production is associated with great leaf surface or a large crown. Good moisture and humus conditions must keep the soil thoroughly moist. Such trees must stand far apart so that they may develop like shade trees. The small crown and the long clear bole of the "lumber tree" are not wanted. Many "sugar bushes," however, are said to be in poor condition due to overgrazing by cattle, which injures the trees and prevents reproduction. Many trees are being overtapped. In some instances, bushes could be improved by removing other species associated with the maples, giving the latter a better chance to develop.

There is a certain amount of romance to the maple-sugar industry, carried on first in a very crude way by the Indians and continued by generations of white settlers in our great north country. Never an industry of large proportions, it always has been one of much local importance. The work comes during a slack time of the year, and the crop does not have to be planted and cultivated. Tapping is a simple operation, the whole family can help, and it yields good cash returns. It is one of those farm "chores" that have greatly enriched rural life. Many a city dweller who was raised on a farm in the North gets the urge in early March to pay a visit to the sugar bush on the old ancestral farm, help harvest the sweetest of all crops, and climax the trip with an old-fashioned "sugaring-off" party.

As late winter days begin to lengthen and the snows get soggy in the daytime, while freezing hard at night, the northern farmer knows it is time to get out his sugaring utensils and start work in the bush. Though methods have changed somewhat, the practice of our forefathers is still essentially in use by the majority of small producers.

The start is made with a team of oxen and a stoneboat or a bobsled, depending on weather conditions, loaded with all the paraphernalia. A brace and a bit is used to bore from one to four holes in each tree, depending on its size. Into these are placed the spigots or sap-drippers, made either of sumac wood or of metal. Pails are hung under these to catch the sap, and daily trips are made with the stoneboat, which gathers up the sap over a wide area. As it is collected from the pails it is placed in a barrel on the boat and drawn to a central location where the sap is boiled over huge open fires in iron or copper kettles. The large commercial sugar orchards are now using the large modern evaporators which can handle several hundred gallons and convert it into syrup in an hour.

Plenty of well-seasoned wood has probably been stored in the woods where the sugaring is to be done. The fire needs constant attention under the steaming kettles. Day and night this refining goes on until the syrup is of just the right specific gravity. It is then ready for further refining in the farm kitchen when it is heated in smaller kettles and pans until the consistency of syrup is reached, or further boiled to make maple sugar. The syrup is placed in one-gallon tins and the sugar cut into blocks or made into various sugar products seen on the market.

Some time near the end of the period, a "sugaring-off" party winds up the weeks of arduous work. The young folks dip simmering syrup from a pan on the kitchen stove, and the liquid is stirred in a saucer until it forms a soft creamy sugar. Or the thick hot syrup is poured upon clean snow, after which it forms a chewy and delicious taffy. Everybody gets all stuck up—but a good time is had by all.

Naval Stores and Tanning Materials

Naval stores are made from the resinous sap of some of our southern pines. In the days of wooden ships, tar and pitch were more important than they are now. Turpentine is still used, however, in the paint and varnish industry, and rosin is used in the manufacture of paper, soap, printer's ink, and other products. The southern states produce more than half the world's supply of naval stores and it is still one of the South's important industries, yielding raw materials worth $50,000,000 annually.

The northeastern, southeastern, and Pacific Coast states produce great quantities of forest products used in the tanning industry. Each year 100,000 tons of vegetable tannins are used, worth $10,000,000, of which about half is derived from the forests. Our native chestnut, though ravaged by the bark disease, still furnishes about one-third of the vegetable tanning materials grown. Both living and dead wood are utilized for their tannin content. The tree formerly ranged from central New England and southern Michigan to northern Georgia, but at least 80 per cent of the trees are now either dead or infested with the blight. This destruction means the loss of an important and easily available source of tanning materials. If blight-resistant chestnuts could be produced and planted on a sixty-year rotation, it would be a profitable crop for the region in which it grows.

The most important source of tanning material, however, is hemlock bark. In the hemlock region of the northeastern states, the original hemlock was felled for its bark and the huge logs allowed to rot in the woods. Now the production of hemlock bark is carried on in connection with the pulpwood and fiber-board industries. If the chestnut supply fails, there seems to be ample hemlock to meet the needs. The eastern hemlock ranges from New England to Wisconsin, and south through the Appalachian Mountains. There remain about 12,000,000,000 board feet. Western hemlock ranges from Alaska to northern California and east to western Montana. The estimated stand in the United States proper is about 85,000,000,000 board feet. As a source of tanbark it is somewhat inferior to the eastern hemlock on account of its thin bark.

Minor Forest Products

In the total aggregate these minor products are not of great financial importance. Locally, however, the rural dwellers who gather them derive considerable income, often when other work is slack.

Sargent's "Manual" lists about 1,200 species and varieties of trees indigenous to our 600,000,000 acres of forest land. In addition, there are some 25,000 species of flowering plants, ferns, and mosses, to say nothing of lesser forms of vegetation such as lichens, algae, and fungi. Of these, certain medicinal plants, those bearing nuts or fruits, and the ornamental trees and shrubs are of economic interest.

The bark of the cascara buckthorn, a tree growing on the West Coast, is an important source of cascara, a medicine useful for its tonic and laxative properties. Gathering of the bark in the spring is an important local industry in Oregon and Washington, west of the Cascades. The cut of bark on six National Forests in this region, over a period of thirteen years, has amounted to about 316 tons, valued at over $30,000. An average of thirty tons of cascara is exported from the port of Vancouver annually.

American forests produce edible nuts and seeds in enormous quantities. In 1929 the pecan crop of the United States was over 25,000,000 pounds, half of which was gathered from wild trees. The pinyon industry of the Southwest produces large quantities of pinyon nuts. In 1925 New Mexico alone shipped out 1,200,000 pounds of shelled nuts. Native Mexicans and Indians gather most of the crop. Edible wild fruits are also locally important. Wild blueberry and huckleberry picking furnishes seasonal employment to thousands. It is impossible to estimate the total amount or value of the fruit picked annually for the table and for the canning industry. Thirty species of blueberry and five of huckleberry are quite widely distributed over the country.

Another minor industry difficult to appraise from a money-value standpoint is that of gathering forest ornamental plants and shrubs. The plants taken vary with the section of the country. In some of the National Forests of the southeastern states such plants are removed under permit for nominal sums. Rhododendrons, mountain laurels, azaleas, and others are shipped out in carload lots from the mountain forests of North Carolina. The proper utilization of this resource must be in the nature of thinning out these plants where they are too dense, rather than a close cropping that would lead to local extermination.

The Christmas tree industry annually uses about 10,000,000 trees. No other country has such a wealth of evergreen trees as the United States, and each section of the country has its favorite species. On the West Coast it may be Douglas fir, spruce, or one of the native firs; in the Northeast it may be balsam fir, spruce, or hemlock. This industry can easily be made a part of good forestry practice. When properly conducted it involves a selective thinning of the young stands annually so as to insure a continuous supply. The Pike National Forest furnishes the city of Denver with about 40,000 Christmas trees and some 500 tons of boughs annually from the dense stands of Douglas fir which profit by a much-needed thinning. The planting of quick-growing European spruce offers opportunities for permanent business and investment.

The cutting of Christmas trees and other decorative materials in New England usually starts in September. In general there is a widespread demand for labor to gather these evergreen materials, months before the Christmas season is at hand. Conifers, ground pine (lycopodium), mosses, mistletoe, kalmia, holly, and other evergreens are gathered in large quantities. Delaware and Maryland are the present center of production of holly greens. In 1927 the holly-products industry of Delaware shipped out 1,500,000 wreaths valued at about $400,000. The holly was harvested chiefly by local farmers.

In addition to these minor products, there are many other edible and medicinal plants—bulbs, tubers, mushrooms, and "greens." There are fiber-producing plants, honey-producing plants, and plants used for basket-making. Others are used for dyes, and for upholstery material; sphagnum moss has even been used for surgical dressings and for horticultural packing.

At least four important facts stand out when considering the protection and the perpetuation of these minor industries:

1. An inventory is necessary to determine the extent and value of these minor forest products.

2. Research must be carried on to determine their beneficial or detrimental effects upon other products and services rendered by the forest.

3. Research must determine the best methods of utilization to put them on a continuous production basis, if that is compatible with other phases of forest management.

4. The object of managing this crop must be predicated on use and not non-use or abuse. No doubt federal and state legislation is needed to protect valuable species. Such laws have already been enacted in New York, Maryland, California, and other states.

We have already noted the use of wood as a source of plastics, textile fibers, and chemicals. Recent experiments have shown that wood can be converted into an endless array of by-products by the use of heat and pressure with the addition of resins, glues, or chemicals. Some of these promise to make wood a real competitor of metals. *Compreg* is a dense, hard substance made of solid pieces of wood or of layers bonded with resin and compressed to one-third its original bulk. It rivals steel in strength. It resists water, alcohol, acids, decay, and termites. It permits the use of woods usually considered too soft for most industrial uses. It is being used to make airplane propellers. *Papreg* is a compressed, resin-impregnated paper made from wood pulp, which rivals aluminum in lightness and tensile strength. It can be made into tubing. *Uralloy* is made of green timber that has been saturated with urea. It can be bent into any shape and permanently set in its new shape by the addition of formaldehyde. It is useful in making curved objects like ships' keels. *Staypak* is a wood product even tougher than compreg, made by heat and pressure. Soft woods like aspen and alder take on an iron-like toughness.

Cellulose, the main component of wood, is already the basis of more than ten thousand chemicals. An important development is the conversion of wood cellulose into sugar by the use of sulphuric acid and the conversion of the sugar into ethyl alcohol. This is made from sawdust and other wood waste. This alcohol is displacing grain alcohol for many industrial uses, as in synthetic rubber. It could also be used in high-test motor fuel. A high-protein yeast for food can be obtained by growing yeast on wood sugar to which nitrogen has been added. Finally, lignin, the cementing material that binds the cellulose fibers in wood, and has heretofore been considered waste, has been made to yield new chemicals and plastics.

CHAPTER IX

THE FOREST AS A LIVESTOCK RANGE

THE HERBACEOUS and woody plants that grow under the forests, in openings between the trees, and in the larger areas that have been cut over comprise one of the major resources of our forest lands. Collectively this vegetation is known as forage and is used by both domesticated animals and game animals. Multiple-use management seeks to obtain the greatest annual returns from each acre of our forest lands, and hence the proper utilization of the forage is important. At the same time its use must be in harmony with all the other uses and services rendered by the forests. This forage plays an important rôle in the production of the country's domestic meat and wool supply, and at the same time furnishes a livelihood to those whose cattle and sheep graze it. It also supports a large and valuable wild-life resource.

Of the 600,000,000 acres classified as forest land (both commercial and non-commercial), over one-half is used for grazing purposes. In the western states over 67 per cent of the forest land is grazed, in the southern states over 69 per cent, and in the central, middle Atlantic, Lake, and New England states about 57 per cent of the farm woodlots are grazed but only about 5 per cent of the large commercial holdings. What this means in relation to the entire livestock industry can be surmised from the fact that about 40 per cent of all the mature range cattle in the western states graze on forest lands for at least a part of the year.

In 1931 when prices were comparatively low, the value of the cattle and sheep grazed on all western forest ranges was about $150,000,000. If complete statistics were available for the whole country, we would no doubt find the forest range lands make up a very substantial portion of our total range lands, and that these constitute a major natural resource.

From the standpoint of National Forest management also, forest range lands are important. It is significant that of the total income from our National Forests, about $7,000,000 per year, usually about 25 per cent, is derived from grazing fees. Fees are paid for approximately 5,700,000 sheep and goats and 1,500,000 cattle, horses, and swine. In addition, there are 5,000,000 head of young animals not counted or charged for. The use of this forage resource by domestic stock, and to a lesser degree by wild life, has a very important bearing on the management of these lands. Forest range management has at least four distinct aspects:

1. As it relates to annual revenue.
2. As it affects the reproduction of timber.
3. As it bears on the efficiency of forest land for watershed protection.
4. As it affects protection from fire.

Aside from other considerations, there is little question that the income from grazing is part of the plan to utilize completely the resources of forest lands. The income from government forest range is important to the states in which these lands lie, because 25 per cent is returned to them for roads and schools, in lieu of taxes not levied. It is also obvious that this income can be constantly increased by careful management. Also there comes up the perplexing question on almost every forest area as to whether it should be managed primarily for forage or for timber. Grazing, especially over-grazing, may be very detrimental to forest reproduction. It may have to be excluded from areas in process of regeneration. On sloping land that is over-grazed, water run-off and consequent erosion may increase to such an extent that serious damage to basic land values may result. On the other hand, the grazing of herbaceous forage under forests is often an absolute necessity to prevent it from becoming a fire menace when dry.

Forest land grazing management in the United States is varied to meet the natural and economic conditions of three large areas, respectively:

1. Western Forest Ranges, constituting those states entirely west of the 100th meridian and South Dakota.
2. Forest Ranges of the South, extending from Texas and Oklahoma and Arkansas through the Atlantic and Gulf Coastal states to Virginia.
3. Forest grazing land in the central, Lake, middle Atlantic and New England states.

In the West, the general problem is the utilization of vast areas of public land by private owners of ranches and livestock. In the South, grazing problems center around the utilization of private forest areas, in both industrial and farm ownership. In the northeastern part of the country, the grazing problem is concerned largely with small woodlots on farms which are used for livestock. Each region has more or less distinct range and forest conditions to say nothing of social, economic, and political conditions, so that for our purpose each region much be considered a separate unit.

Western Grazing Lands

In order to appreciate the great changes that have taken place in the ownership, distribution, and carrying capacity of western grazing lands it will be necessary to go back into American history and fix in our minds a few important dates and events. The part played by the cattlemen in opening up and developing the great West is an epic full of romance and adventure, and one of the most fascinating periods of our expansion as a nation.

After the Peace Treaty of 1783 the first large acquisition to our original colonies was the Louisiana Purchase in 1803, and the last one was the Gadsden Purchase of 1853. These vast acquisitions, comprising about 1,400,000,000 acres, were to be known as the Public Domain. For over fifty years after its acquisition, this Public Domain was to play a major rôle in the economic and financial development of the West. Today only about 165,000,000 acres of unreserved Public Domain is left.

Among the first pioneers who dared to explore the trackless wilderness were Lewis and Clark in 1804. They crossed the Rocky Mountains and the next year reached Oregon by way of the Columbia River. In the same year Zebulon Pike explored parts of the Rocky Mountains. Reports of what these men found started a steady migration westward. Soon the great lines of travel were marked out as the Oregon, Santa Fé, and Overland trails. These were soon worn deep by the covered wagon and pack trains which carried the pioneers westward to the rich northwest, the gold fields of California and Colorado. That great empire of grass—the prairies and plains—even as late as 1865 was still a wilderness known only to the Indian and the buffalo. An area 1,500 miles long and 800 miles wide stretched from the Mississippi to the Rockies and from Mexico to Canada. One billion acres of luxuriant grass was waiting for anyone who would come and take it.

Following the Civil War, two more or less antagonistic movements started. One was the opening of the great grasslands to settlers by the new railroads and the other was the coming of the cattlemen from Texas with their longhorns to take possession of the great range country. Plowmen from the East came to "turn the sod wrong side up," aided and abetted by government land grants and the new railroads. Texas longhorns began to pour into Kansas, and later Wyoming and the North, over the famous Chisholm trail—and they were destined to become the foundation stock of large northern herds financed by eastern and foreign capital. Sooner or later these two interests were bound to clash.

Let us look at the agricultural exodus to these new parts. In the late sixties the great railroads were given huge land grants and other government help to bring civilization to the West. The Union Pacific was finished in 1869; the Northern Pacific not until 1883. In addition, a most liberal land policy by the government did everything it could to give away the land to anyone who would settle on it. In 1862 the Homestead Act would give 160 acres; later by another act 640 acres were offered to anyone who would comply with certain occupancy regulations. Then came the desert land laws in 1877 and the Timber and Stone Act in 1878, with other land and mining laws all aimed to develop the western empire.

Thus between 1860 and 1890 the great iron horses rapidly accelerated the migrations started by the covered wagons and pack trains. A thin trickle of settlers and homesteaders dotted the railroad right of way across Kansas and Nebraska in the early seventies, and this was soon augmented by the march of the cattlemen from the south to deliver their herds to the railheads to be shipped east. Abilene, Newton, and Wichita were soon important cow towns, to be followed later by Hays City, Dodge City, and others. Of course the more homesteaders came the greater was the inroad on the grassland. Sod-house homesteads, corrals, and barbed-wire fences sooner or later mean the end of open ranges. It meant that the cattlemen would seek the open grasslands to the north—Nebraska, the Dakotas, Wyoming, Montana, and even Canada. Up to 1877, however, the Indians had something to say about this.

During Civil War days Texas had accumulated a great surplus of cattle. After the war these were destined to pour out over the western prairies and plains. It was about 1867 that the first herd of Texas longhorns was driven over what was later to be known as the Chisholm trail to the railheads of Kansas. In the next twenty years about 10,000,000 head of steers were destined to go north. Various feeder trails around San Antonio, Texas, joined to form the main trail which lead through Oklahoma, Kansas, and Nebraska, and later into Wyoming and points north. Herds followed the grass north in the spring, and so luscious was the grass that despite the long trek of 1,500 or more miles the longhorn steers would add on the average of 200 pounds to their weight by the time they reached their destination.

These great treks north were in reality great adventures, and many stories are extant of the hardships encountered. Late blizzards, swollen rivers, thunderstorms, stampedes, and wild Indians made these trips hazardous. The rate of travel was about twelve to fifteen miles a day. Wood for camping was scarce except along the rivers. On the open prairies wood was carried in the chuck wagon, and there were also "buffalo chips" for fuel. In general there were good water holes en route, and the river fords were mostly safe and well marked out. Once Texas was left behind, encounters with the Indians were common. Herds were stampeded, cowboys killed, and steers lost. At best the Indian warriors would demand their tribute of meat, tobacco, or whatever their fancy demanded. They pleaded, justly, that since the white man was killing off the buffalo, they and their children were starving.

The Indian disputed every foot of the grassland country with the white man. The pressure of mere numbers was destined to spell the doom of the Indian. The white man was also relentlessly slaughtering the buffalo, which furnished the red man with food, shelter, and clothing. It is said that about 1875 there were thousands of professional hunters killing the buffalo merely for their hides. Gradually the red man was pushed back. The Cherokees, Cheyennes, Sioux, Osages, Pawnees, Comanches, Arapahoes, Wichitas, and many other tribes were gradually forced to take to reservations allotted to them and leave the grasslands to the cattlemen and the homesteader.

The final act in this western drama came, of course, in June of 1876 when General Custer and his Seventh Cavalry were massacred almost to a man by the rebellious Sioux, commanded by Sitting Bull. This happened in southern Montana. As a direct result the Sioux were herded into their reservations the next year, and the cattlemen prepared to invade the country north of the North Fork of the Platte River in Wyoming. With the Indians subdued, the plowman and the cattlemen had the range from Canada to Mexico to themselves.

The Golden Age for the pioneer cowman was said to be about 1884-85. Million-dollar cattle outfits were in existence even at this early date. But a few years later the end was in sight. The bad winter of 1886-87 killed much stock, and many companies went broke. More railroads came, and with them settlers, homesteaders, government surveyors, and of course more barbed wire. Later came sheep from the South, and in 1891 the law creating forest reservations. All these factors together spelled the doom of open free range and meant that the time was not far distant when small outfits, allotted range, greater care in winter, and greater all-round efficiency were to be the order of the day. Finally, in many localities overgrazing had so reduced the carrying capacity of the land that it was very evident that government regulation was the only thing which could save both the forage resource and the industry.

Government Control

An important change in the status of western grazing lands was made when the Law of March 3, 1891, was passed, giving the President power to establish forest reservations. It was not, however, until 1905 that the Division of Forestry became an administrative organization. Since that date the Secretary of Agriculture has had the authority to permit, regulate, or prohibit grazing on the Forest Reserves, later National Forests, the policy being to allow as much grazing as the proper care and protection of the forest and water supply would permit.

Grazing use of National Forest land is today considered a personal and non-transferable privilege. It is on a permit basis because it is allowable only so long as it does not interfere with the purposes for which the National Forests were created. It is non-transferable because it is based on the possession of certain qualifications peculiar to the permittee.

Long before the National Forests were created, stockmen had been allowed to graze their animals on public lands. This use, continuing for a long period of years, had, in the absence of congressional legislation, been commonly accepted in many communities and had even been recognized by certain courts of law. Grazing was allowed under "unwritten law" as it were, only by the passive consent of the United States. By force of the presidential proclamations creating National Forests, such passive consent ceased, being superseded by definite regulations under the Secretary of Agriculture as prescribed by the authority of Congress. By this early use of the range, these stockmen were recognized in these localities as having preference rights or equities in the use of these lands. Such equities form the basis upon which grazing privileges are allowed. Prior use and occupancy of National Forest lands for grazing purposes is the first and foremost requirement. Local residence and ownership of improved ranch property within or near the Forest and dependence upon government

U. S. Forest Service Photo

ROCKY MOUNTAIN GOAT, CHELAN NATIONAL FOREST, WASHINGTON

BIGHORN SHEEP, JASPER NATIONAL PARK, ALBERTA, CANADA

Photo Canadian Travel Bureau

Photo Department of the Interior, Canada

ANTELOPE, BANFF PARK, ALBERTA, CANADA

YOUNG FAWNS BEING FED, PISGAH NATIONAL FOREST, NORTH CAROLINA

U. S. Forest Service Photo

Photo New York Conservation Department

BEAVER IN THE ADIRONDACK MOUNTAINS

A BEAVER DAM, NEW YORK STATE

Under protection, the beaver is increasing in New York and Pennsylvania.

Photo New York Conservation Department

U. S. Forest Service Pho

FISHING AT BASE OF MT. ADAMS, COLUMBIA NATIONAL FOREST, WASHINGT

FISHING ON THE RESTIGOUCHE RIVER, NEW BRUNSWICK, CANADA

Photo Canadian Travel Bure

Photo courtesy Maine Development Commission

STRIPPING FISH OF EGGS AND MILT AT A MAINE HATCHERY

FISH HATCHERY, WARRENSBURG, NEW YORK

Photo New York Conservation Department

U. S. Forest Service Photo

BASS FINGERLINGS GO BY PLANE, SUPERIOR NATIONAL FOREST, MINNESOTA

YOUNG TROUT HATCHING FROM EGGS AT NEW YORK STATE FISH HATCHERY

Photo New York Conservation Department

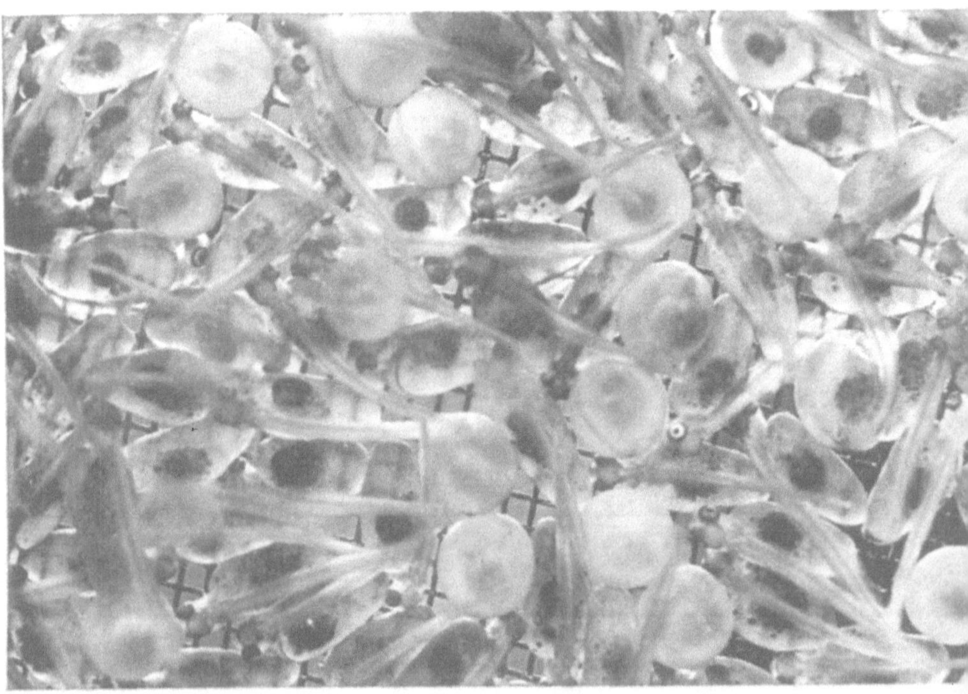

Photo by Bayard Wootten

FEEDER FOR THE TVA

Trees hold mountain soils, protect watersheds, and prevent floods and erosion. Little Tennessee River near its source in the mountains of North Carolina.

Photo U. S. Soil Conservation Service

SILT-FILLED RESERVOIR, SPARTANBURG, SOUTH CAROLINA

Soil washed from near-by farms choked this reservoir within two years.

STORAGE CAPACITY RUINED IN THIRTY YEARS

This view of reservoir at Schoolfield, Virginia, shows how abuse of forest and farm land upstream affects power-plant and water systems downstream.

Photo U. S. Soil Conservation Service

Photo U. S. Soil Conservation Service

BEFORE: Eroded field in Jackson County, Illinois, planted with black locust trees in 1934.

AFTER: Same field three years later. Since black locust grows readily on poor soil, it is a valuable tree for reclaiming eroded land.

Photo U. S. Soil Conservation Service

Photo U. S. Soil Conservation Servi

TREE-ROOTS EXPOSED BY EROSION, WASHINGTON COUNTY, NEBRASKA

Over-cutting, over-grazing, or repeated forest fires can produce this condition. As the root-system of the tree is exposed, its nourishment is cut off.

ERODED FIELD IN SUMTER NATIONAL FOREST, SOUTH CAROLINA

As the soil is destroyed, wealth and living standards decline. The landscape here was once beautiful and the soil fertile—but the dilapidated cabin suggests the extent of "human erosion." Such land should be returned to forest.

Photo U. S. Forest Servi

Photo U. S. Soil Conservation Service

EROSION HALTED BY KUDZU VINE

The very existence of this farm in Marion County, Georgia, was threatened by the gnawing cancer of erosion. Kudzu vine, like the black locust tree, is a leguminous plant which enriches poor soils by drawing nitrogen from the air.

STRIP-CROPPING AND TERRACING IN THE UPLAND SOUTH

This farmer is saving his land, even though the slope is heavy. The numerous patches of woods also help control the run-off of water.

Photo U. S. Soil Conservation Service

Photo U. S. Soil Conservation Servi

BRIDGE WRECKED BY FLOOD, CAMBRIA COUNTY, PENNSYLVANIA

DRIVEN OUT OF HOUSE AND HOME

Throughout the country there is a loss of over $110,000,000 a year from floods, not to mention the large loss of life.

Photo U. S. Soil Conservation Servi

Photo U. S. Soil Conservation Service

FLOOD IN WYNOOCHE VALLEY, WASHINGTON

U. S. Forest Service Photo

COLLECTING MAPLE SAP IN VERMONT

Vermont is a leading state in the production of maple sugar and syrup. The crop for the entire country is worth $5,000,000 a year.

A GREAT MODERN LUMBER MILL

Forests in background and houses for workers in front. The forest and lumber industries furnish full-time employment for about four million people.

U. S. Forest Service Photo

U. S. Forest Service Photo

BUILDING OAK BARRELS FOR TURPENTINE

More than 4,500 uses of wood have been tabulated, and in spite of substitutes the number is increasing.

SHEET OF CELLOPHANE LEAVING COAGULATING BATH

Photo courtesy E. I. du Pont de Nemours and Company

Photo courtesy E. I. du Pont de Nemours and Company

OBJECTS MOLDED FROM PLASTACELE

range are also conditions that may entitle a stockman to grazing privileges. Stockmen who have acquired by purchase or inheritance stock grazed on National Forest lands under permit may also be granted privileges.

Of the 144,000,000 acres of forest land that are grazed in the western states, 64,000,000 are located within the National Forests. The grazing ranges within these forests are valuable because they contain a very large share of the summer feed for beef cattle, sheep, and range horses of the Rocky Mountains and Pacific Coast states. Just as important is the fact that under government administration, protection from fire and the regulation of grazing has tended to stop deterioration and restore to some extent the forage crop. The same cannot be said of most forest range land in private ownership.

Finally, government forest range lands directly or indirectly affect the prosperity and development of a large part of the West. There are many agricultural communities adjacent to these forest ranges that are dependent for their welfare on the production of livestock. Farms are small, far from markets, and valuable principally for producing hay and other feed crops. Without the aid of the forage in the near-by forest areas for complementary feeding, these thousands of communities would never have reached their present stage of development. Millions of dollars are invested in improved ranch properties, buildings, and water development, and additional millions are spent in leasing additional private grazing lands.

In addition to the forest range lands owned and leased by the federal government, there are forest range lands under other forms of ownership which present distinct problems. There are 42,000,000 acres of private lands, 22,000,000 acres of unreserved public domain, 12,000,000 acres of Indian reservations, and 4,000,000 acres of state lands. It must be borne in mind, however, that there are also vast areas of non-forested, open range lands in the West which are not included in this discussion, because we are dealing primarily with those range lands covered more or less with a forest growth. In the National Forests, for example, besides the 64,000,000 acres of forest range lands there are some 24,000,000 acres of other usable range.

Forest-Range Types

The various western forest types produce forage varying in quantity and quality. The ponderosa pine forest furnishes good feed over large areas on the lower mountain slopes and plateaus. The open character of the stand permits the growth of a great variety of herbs, shrubs, and bunch grasses. It comprises the most extensive and important forest range type. The sugar pine-ponderosa pine forests of California are somewhat more dense than the typical ponderosa pine type, and therefore contain less feed. This is mainly brush, grasses, and weeds. The aspen type of the Rocky Mountains ordinarily grows in rather open stands and usually has a good undergrowth of grasses, weeds, and browse. The woodland or juniper type on the lowest slopes and in the foothills is an important type and furnishes the same kinds of feed.

The lodgepole pine of the Rockies, the Douglas fir, redwood, and the spruce-fir types are usually too dense to allow satisfactory grazing. Only at the higher elevations, in the sub-alpine type, does the spruce-fir forest become open enough to produce good feed. This is an important source of forage in the late summer. The mountain brushlands both in the Rockies and in California and Oregon, as a rule, contain brush species of low forage value.

Western Forest-Range Conditions

In the days of free range, before the National Forests were established, an apparent abundance of feed and the desire for quick profits brought large investments in cattle, and excessive overstocking became general. In the late eighties, after a bad winter and a few severe drought years, the livestock industry suffered a severe setback. Range conditions in general were getting bad. The depletion of the forage, which reached the denudation stage in many areas, the erosion of fertile topsoil with consequent damage to farm lands and irrigation works, the replacement of palatable grasses and herbs by worthless brush and weeds, and the destruction of forest reproduction by half-starved animals, all helped seriously to lower the carrying capacity of all range lands and threaten the stability of the livestock industry.

Forest range lands suffered greatly in these processes, because with the depletion of open range lands below the lower timber line, stockmen sought the higher forest ranges which were usually cool and well-watered. The annual fires set in many parts, in the hope of improving range conditions, often actually destroyed range and timber values and injured the soil.

This, in general, was the situation when the National Forests were created. Since that time much has been accomplished, mainly by fire control measures and by the regulation of grazing. In one area in central Utah, for example, the grazing capacity was four times greater in 1927 than in 1912. Similar improvements are recorded on many National Forest areas of the West. Where grazing is carefully regulated, experiments have shown that not only is the carrying capacity increased, but livestock losses are far less, calf crops are larger, the cost of producing calves is less, and the sale value is correspondingly greater.

On private holdings and on the unreserved public domain, however, the story is different. Unsatisfactory range conditions here are still having a decidedly deleterious effect on the social and economic conditions of the West.

Western Watershed Problems

The loss of productivity of forest range lands means the loss of income and taxable wealth. Often, however, this is only the forerunner of far greater and more serious losses, if watershed protection, timber production, recreation, and wild-life services are involved.

The destruction of the protective cover of herbaceous and shrubby plants is soon followed by the erosion of the litter of decaying vegetable matter and later of the mineral soil itself. Widespread sheet and gully erosion in the western states has robbed these areas of the fertile and productive surface layer which is the basic forest-land resource. As a result, the rapid run-off of water has increased the destructiveness of floods, added greatly to the silting problems of reservoirs and other irrigation works, and has jeopardized the heavy investments in irrigated farm lands, irrigation and power properties, and domestic water supplies. In addition, timber production, recreational facilities, and wild-life services have also suffered.

The public interest involved in only one of these resources is shown, for example, by the extent to which the people of the West are dependent upon National Forest areas for their domestic water supply. About 2,500,000 people living in 700 cities and towns in the western states obtain their municipal water supplies from these areas.

Western Grazing Management

The most important forest grazing problem is the restoration of depleted ranges. This will at the same time enhance watershed, wild-life, recreational, and other forest values. The most feasible means of doing this is to afford the native vegetation an opportunity to reëstablish itself by careful management. This means that important forage species must be allowed to develop sufficient growth to regain their former vigor and that proper conditions must be provided for their reproduction. In extreme cases this might mean the exclusion of all grazing from the worst areas. On favorable sites the sowing of seeds of the better cultivated forage plants and of desirable native range species would speed up the restoration.

Deferred and rotation grazing is one of the several methods of revegetation developed by the Forest Service. This consists of deferring grazing on a part of a range-unit each year until the more important forage plants have formed a vigorous growth and have matured their seeds. The rotation feature comes in through deferring grazing for a year or two on different parts of the unit in succeeding years. By such late-season use the mature seeds are shaken to the ground and partly buried by the trampling of the stock. The next year light or deferred use may be necessary to assist the growth of the new seedling plants. It is estimated that about 900,000 acres are in need of early revegetation by methods similar to these.

There are several principles that must be employed in the handling of livestock on the range if the maximum returns are to be expected from the forage. Ranges should be used by the class of stock best suited to use it. The number of livestock must be adjusted to what the range can support satisfactorily. The season of use must be adjusted to the most satisfactory period so as to derive the greatest value from the use of the feed. Grazing must be distributed over the range in such a way as to reduce damage, to insure uniform utilization and to facilitate the use of feed of special value at certain seasons. Finally, sound forest-range management must consider range improvements such as fences, water developments, cabins, corrals, and driveways, and the control of poisonous plants, rodents, and predatory animals.

Outside of National Forests, the most pressing problems are on the private grazing lands and on the former unreserved public domain. The latter contains about 21,500,000 acres of grazed forest land which is under no regular management. If such range land, together with the millions of acres of non-forested public domain, are to be saved from utter destruction, it is obvious that they must be incorporated into one of the governmental agencies so that management and restoration will be made possible.

The Taylor Grazing Control Act of 1934 had for its purpose the solution of the then unreserved public land problem. About 140,000,000 acres of this land was in 1936 under the administration of this act. It will not only regulate the use, the preservation, and the rehabilitation of a large part of the public domain, but it will also put an end to unwise homesteading under existing laws. It has already put an end to the era of free public range.

Before leaving the discussion of western range lands, it might be well to get an over-all picture of the range, both forested and non-forested. The figures given for western forest range lands must not be confused with the large areas of formerly unreserved and unallotted public range lands of the West having no forests on them. A few further data will make this difference clear. The total livestock range land of the western United States (west of a line drawn from central North Dakota south to the southernmost tip of Texas) is about 728,000,000 acres, or nearly 40 per cent of the total land area of the nation.* This is distributed roughly as follows:

National Forest ranges	88	million	acres
Indian lands	48	"	"
Other public lands (Now mostly under the Taylor Act)	151	"	"
State and County	65	"	"
Private	376	"	"
TOTAL	728	"	"

As will be seen, the National Forest ranges make up only about 12 per cent of the total. Except for the National Forests most of this range has been exploited since Civil War days. The average depletion of this important forage resource is 52 per cent. National Forest ranges show the least depletion, or 30 per cent, while the other public domain (now grazing districts) show the most, namely, 67 per cent.

SOUTHERN FOREST RANGES

It is estimated that three acres out of every four of southern forest land is used for grazing purposes. About 20 per cent of the grazed area is in farm woodlands, about 78 per cent is in large timber and turpentine holdings, and 2 per cent is in public ownership.

Forests of the South furnish feed for livestock, principally cattle, during the spring and early summer. It is essentially the open range type of grazing similar to that in the West. The commercial pine and hardwood forests have been largely cut over and the resulting open stand on the restocking land supports a good growth of grasses, herbs, and shrubs. The lands in general are unfenced (except in certain states and counties where fence laws have been passed) and most of the stock is owned by local residents who seldom own the land on which the grazing takes place. The leasing of grazing rights is not the common practice, since the unfenced forest lands are regarded as public ranges. In the southern pine type large areas of woods are burned annually or periodically by livestock owners, which practice

* *The Western Range* (Sen. Doc. 199, 1936), pp. 7-41.

seriously interferes with the satisfactory regeneration of the forest. A poor grade of livestock and a carefree type of management still prevails on much of the area.

According to the 1930 census, nearly 15,000,000 cattle, 9,000,000 sheep, and over 9,500,000 hogs grazed on all types of range lands in the South and a very large portion of these used forest land. The grazing of livestock on forest land furnishes a livelihood or supplementary income to a large portion of the rural population. Owners of such land also receive an income to meet the carrying charges on the land. Grazing has for many years been one of the South's leading industries—though scattered and unorganized.

A considerable number of forest types offer good grazing in this region. The longleaf-slash pine type, which follows the coast, offers an abundance of "wire-grasses," introduced grasses, and legumes. The shortleaf-loblolly pine-hardwoods type is found inland, and is also rich in botanical species including broomsedges, needle grasses, and legumes and many hardwood species which are browsed. In the southern Appalachians the oak-chestnut-yellow poplar and at low elevations the oak-pine types furnish broadleaf tree browse and numerous shrubs, besides native woodland grasses. Grassy "balds" at an elevation of 5,000 feet or more, covered with redtop and blue grasses, furnish feed of high grazing value. In Texas and Oklahoma, where the forests are more open and scattered, there is considerable hardwood sprout browse and various nutritious grasses.

The most important grazing problems of the southern forests are: the general practice of uncontrolled burning allegedly to remove dry grass, to keep down undergrowth, to control insect pests, to make hunting better, or to improve the quality of the forage; the serious damage done by hogs in rooting out longleaf pine seedlings; the grazing of the sprouts of valuable hardwoods; and the need of coördination between forest-range grazing, the use of improved pastures, and supplemental winter feeding.

No doubt uncontrolled burning is the supreme conservation question of the South. An average of 37,500,000 acres of forest land were burned annually from 1926 to 1930 in the South. Forest Service studies have shown the great damage to the southern pine forests, especially to turpentined trees, by these fires. Forest reproduction on large areas is made impossible, especially if the uncontrolled burning is at too frequent intervals and at the wrong season of the year. The coördination of grazing with timber production requires either the prevention or the control of fire to insure sustained yield forestry. This conflict of economic interest must be worked out on each acre on the basis of the greatest good to the greatest number in the long run. This subject is treated more fully in Chapter XVI.

Forest Grazing in the Northeastern Section

The grazing on forest lands in this region is confined almost entirely to farm woodlands. It is principally the pasture type of grazing as contrasted to the open range type of the West and South. Of the 63,000,000 acres of farm woodland, 36,000,000 acres (1930 census) are grazed. In addition there are about 6,000,000 acres of forest land in other private ownership that are grazed.

The grazed forests are important to the social and economic welfare of the region on account of the wood products which they furnish, and their high watershed-protective value. Their recreational and wild-life values are essential to the health and happiness of both the urban and the rural population of the great farm belt. By the 1930 census the region had a total of about 95,000,000 head of cattle, sheep, and hogs, about 70 per cent of which were in the central states. A considerable portion of the livestock uses forests at some time of the year.

In general, little real forage is produced under a well-stocked forest. Since, however, most of the forests of the region have been cut over, the open stand produces tender shoots of hardwood trees and shrubs. Mast from oaks, hickories, walnuts, beech, chestnut, and other trees are widely used by hogs. With continued cutting and grazing, grasses appear in the openings, both native and naturalized introduced species. With continued heavy grazing the valuable grasses give way to less useful grasses and weeds.

In this region there is urgent need for coördination between grazing and timber production, especially in parts of the central states where investigations show that the woodland pastures are being used by at least five times as many livestock as the grazing capacity would support, indicating heavy damage to forest reproduction. A thoroughgoing appraisal of grazing and timber values is needed to determine which woodlots are to be converted into pastures and which are to be used solely for forestry purposes. Grazing of farm woods in the northeastern and central hardwood regions is definitely considered detrimental to the woods, and as compared with improved pasture, such woodlands furnish relatively poor grazing.

CHAPTER X

FORESTRY AND AGRICULTURE

FORESTRY IS RELATED to agriculture in many important ways. First of all, farm woodlands are a major source of forest products. Agricultural economists class forest products as a major farm crop, since it usually ranks within the first ten farm crops in point of value. The farmer is not only a producer of forest products but he is often engaged in the manufacture of them when farm work is slack. The farmer is likewise one of the greatest consumers of forest products. For him there is no cheaper and better raw material for building purposes than wood. The intimate relation of forests to water supply directly affects his domestic water supply, his irrigation water, cheap water transportation on rivers, cheap hydro-electric power, and the stability of his investment by protecting land against erosion and floods. Bird life of the forest helps to control insect pests of farm crops. Forests are an important source of meat crops, especially in the western states where millions of head of livestock graze on forest ranges. Finally, the increasing use of the forest for hunting, fishing, and other recreational uses brings direct cash returns to the rural population.

Of the 6,800,000 farms listed by the 1935 census, about two-thirds had farm woodlots. The total area of these was 185,000,000 acres, comprising an area equal to about one-third of our total forest land and about one-half of our cultivated land.* An area of such tremendous aggregate size must command attention in any scheme of planned land economy.

Practically every farm has woodland, provided the climate is favorable to tree growth. This woodland is usually the part of the farm that is too rocky or too steep to permit cultivation. Farm forests usually require no planting or cultivation, and provided they are properly managed and fire and grazing damage is kept under con-

* *Living in Forest Lands* (U. S. Forest Service, 1938), p. 19.

trol, they usually reproduce well and yield an abundance of materials. These can be harvested at a time of the year when regular farm work is slack. Finally, as the large centers of lumber production become depleted of virgin timber, the wood-using industries will turn more and more to the farmer for their raw materials. They have already done this to a considerable degree in the great hardwood belt of the Middle West, which is favorably situated near a great wood-using market.

A few figures will indicate the magnitude of the farm woodland business. Farm forest products in the United States in 1929 were estimated by the Bureau of the Census to be worth over $240,000,000. They consisted largely of the following items:

34 million cords of firewood.
1½ million cords of pulpwood.
100 million fence posts.
15 million railroad ties.
3 million poles and piling.
5 billion board feet of saw logs and veneer logs.

The states which show the largest income from this source are Wisconsin, New York, North Carolina, Tennessee, Virginia, Missouri, and Minnesota.* In fact it is surprising to note that one-third of the timber of all kinds that is cut annually comes from farm woodlands.†

It follows from these figures that the farmer is no small lumberman himself. He adds to farm income not only by selling his products, but also by selling his labor. He finds profitable employment not only for himself but for his team and equipment after his farm crops have been harvested. He often works in near-by logging camps or mills. He may sell his wood crop on the stump and let others do the cutting and converting, but the farmer who selects and cuts his own timber and grades his own logs and converts them into finished products, reaps the greatest profit. With the advent of power farming, portable saw-mills have come into wide use with a consequent increase in products cut and sold. He may cut his own trees and also those of his neighbors. The timber may be cut into ties, fence posts, and poles, or into shooks and bolts for barrel staves and implement handles. Thus it is evident that the farmer can derive a continuous source of income from his woodland by careful management to insure reproduction.

* *A National Plan for American Forestry* (Sen. Doc. 12, 73rd Cong., 1st Sess., Washington, D. C., 1933), p. 897.

† *Living in Forest Lands* (U. S. Forest Service, 1938), p. 20.

In fact, the entire forestry program—national, state, and local—helps the farmer tremendously in providing raw materials for wood-using industries. The return of many of the small wood-converting operations which once were abundant when virgin timber was available, will not only bring employment and payrolls and better markets, but will indirectly increase the tax base, lower farm taxes, and provide for more and better roads and schools. This would revitalize our rural economy.

The many excellent physical properties of wood make it indispensable for farm use. The replacement and repair of wooden buildings, fence posts, poles, and other farm equipment creates a steady demand for wood in various forms. Wooden boxes, barrels, and crates are consumed in large quantities for marketing farm crops. The farmer is the greatest user of fuelwood, and what is more he obtains it at low cost.

The Farmer's Water Supply

The rôle of forests, often those far removed from the farm, in regulating water run-off and checking erosion is of greatest service to the farmer. By holding the soil in place on steep slopes and along stream banks the continuity of agricultural enterprise in the valleys is assured. In regions where these slopes have been denuded of their growth, thousands of acres of rich bottomlands have been ruined by deposits of sand, gravel, and boulders. By the filling up of power dams with this debris the farmer is deprived of cheap electric current to run his farm. Hyrdo-electric power development requires a steady flow of clean water. Rivers whose flow is regular offer the best cheap water transportation for farm products. Also the wells and springs on the farm are in no small degree dependent on the near-by forest cover, for this cover determines the amount of water that seeps into the ground to feed these sources of water supply and also determines the amount that is lost by rapid surface run-off.

Speaking of water supply for the farmer, more than 240,000 farms, or about half the total in the eleven western states, are dependent upon irrigation for watering their crops. These irrigated lands have a combined area of 17,500,000 acres and the total value of lands, buildings, and machinery on these farms is about $4,500,000,000. Nearly another billion dollars is invested in reservoirs and distributing systems. The majority of western streams that supply this irrigation water rise in the National Forests located at their headwaters. Without these protection-forests the flow of water would be irregular and uncertain, larger and more costly storage reservoirs would be needed, and great losses in silting of reservoirs and ditches would result.

BIRDS AND AGRICULTURE

The bird life of the forest is one of the farmer's greatest assets. The financial damage done by predatory insects and the estimated value of birds in checking these has already been mentioned in the chapter on wild life. At this point we shall add some details on the relation of birds to agriculture.*

We must be reminded that a number of factors contribute to the elimination of bird life in the forest. Chief among these is the destruction of forests, which removes the nesting places of many of the woodland species. This destruction has also affected the water economy of large areas, whereby destructive floods, erosion, the silting of streams and lakes, and the lowering of the water table has destroyed many nesting places of shore birds, ducks, and other semi-aquatic species. The destruction of native trees and shrubs which bear fruits has been a serious factor in the decimation of bird life. Finally, as advancing civilization has encroached upon bird habitats, a number of enemies are menacing bird life, the chief of these being the domestic cat, which is allowed to roam the near-by thickets and meadows unmolested—and cats destroy far more birds than is generally realized.

The remedy for such a situation is clearly more and better forests managed on a continuous production basis. The restocking of decimated woodlands and forest borders, the reforestation of bare, waste lands, and the reëstablishment of our forest-water economy would be most effective in increasing our native bird population. Secondly, the establishment of bird sanctuaries—local, state, and federal—must be encouraged. And then, too, there are the birdhouses, winter feeding, and the control of stray cats.

Specifically, how do birds aid the farmer? Chiefly in three ways: they eat insect pests which destroy his crops, they destroy quantities of weed seeds, and they destroy rodents and other undesirable farm pests. While it is conceded that a few species destroy cultivated grain, fruit crops, and garden produce, and may even prey on useful animals, it is fully demonstrated that the benefits of bird life far outweigh the damage done. For example, the hawks as a group are often regarded as destructive by the farmer. Unquestionably the Cooper's hawk and the sharp-shinned hawk are injurious because of their preying on poultry and wild birds—but the broad-winged hawk, the

* See J. Henderson, *The Practical Value of Birds*, New York, 1927.

rough-legged hawk, and the sparrow hawk devour a great many grasshoppers, beetles, crickets, caterpillars, mice, gophers, and other rodent pests and hence must be regarded as helpful to the farmer. Even the starling and the English sparrow, two so-called pests with evil reputations, have been found to be valuable, because they destroy the Japanese beetle and the Colorado potato beetle. These birds also eat weed seeds.

The birds that are most beneficial in the destruction of farm pests are the cuckoos, woodpeckers, king birds, crows, jays, orioles, finches, sparrows, tanagers, vireos, warblers, thrushes, wrens, nuthatches, and chickadees. Birds are exceedingly active, and for their size consume a large quantity of food. E. P. Felt, formerly State Entomologist of New York, records many observations bearing on this. A pair of tanagers, for example, was observed to eat thirty-five newly hatched caterpillars in a minute and to continue at that rate for eighteen minutes. Two kingbirds killed 250 gypsy moths in three hours. Two Maryland yellowthroats destroyed about 7,000 (estimated) plant lice in one hour. The study of the contents of bird stomachs by federal and state agricultural departments throws added light on bird diets. A nighthawk's stomach was found to contain the remains of over 300 grasshoppers and over 50 beetles, bugs, and similar pests. A killdeer's stomach was found to contain 300 mosquito larvae. Bank and cliff swallows show as high as 68 boll weevils in their stomachs. The crop of a ring-necked pheasant contained 8,000 seeds of chickweed. Mourning doves contained on the average of 7,000 weed seeds per bird.

All of which adds up to the conclusion that if we expect to control farm pests which are dangerously menacing our agricultural crops, we must rehabilitate our vast forests with their abundant species of bird life.

Farm Woodland Pasture

Farm woodland pasture is a source of food for livestock owned by farmers. This is especially true in the eastern and central parts of the country, since in the West the summer range is mostly in the remote mountain areas and in the South it is mostly under the extensive pine forests. The grasses, weeds, and shrubs under the trees and in open spaces in the woods are an important farm resource. The mast of oaks, hickories, walnuts, beech, and chestnut are used by hogs. This constitutes the pasture type of grazing as opposed to the open range type found in the South and West. In the area embracing the central, Lake, middle Atlantic and New England states, there are 63,000,000 acres of farm woodland, of which over 36,000,000 are grazed. There are perhaps in addition some 6,000,000 acres of forest land in other private ownership that are grazed in this region. According to the 1930 census these regions have a total of 36,500,000 cattle, both beef and dairy breeds, 17,500,000 sheep, and 41,500,000 hogs. By far the larger part of these animals are, of course, in the central states, and a large part of them graze on farm woodlots.

Forest pasturing in the central states, as well as in other sections, is due for considerable modification if the woodlands are to be spared. Three out of every four farm woodlands in the central states are grazed. The forest-grown forage on the whole is of comparatively little value, but grazing destroys forest reproduction and reduces watershed benefits. The exclusion of livestock from these woodlands, except the small areas needed for shade, would not seriously interfere with livestock production, but would greatly benefit the forests and increase farm-forest income.

The Farmer's Benefit from Forest Recreation

It is certain that a very large portion of the five billion dollars that is poured annually into the recreation industries by the American public directly or indirectly aids the farmer. He sells both his labor and his crops, the latter often directly to the consumer at better prices. He can rent bungalows and camp sites. He can take in tourists and boarders. The management of forest wild life for continuity will increase his income from hunters and fishermen. He himself can trap and sell the furs and can hunt food animals and thereby reduce household expenses.

Value of Planted Forests

The protection given by forests and windbreaks to crops, buildings, and livestock against winds is of great value to the farmer. In the Prairie and the Great Plains, where natural growth of trees is usually limited to the banks of streams, forest planting on farms is done mainly for protection. These strips of forest hold snow to increase soil moisture, check wind velocity to reduce drying and erosion of the soil, and greatly improve the appearance of the farmstead. This protection adds up to more and better crops, more milk per cow, and quicker growth of meat animals. Its aggregate value to the farmer is hard to estimate.

Coöperative Aid to Woodland Owners

Much encouragement and financial aid has been given owners of woodlands to insure sound and profitable farm-forestry practices in recent years. Beginning with the Clarke-McNary Act of 1924, a number of coöperative federal-state laws have been passed to assist farmers both in an educational and in a practical way to deal with all phases of woodland management. This includes not only reforestation and management for continuous production, but also operations dealing with harvesting, utilization, and marketing.

The combined brains and resources of the federal and state governments through the land grant colleges, agricultural experiment stations, and forestry schools have been mobilized to allow the county agricultural agents and the extension forestry specialists to give direct, on-the-ground help to farmers. The Clarke-McNary Act provides for direct coöperation with the states in the procurement, production, and distribution of forest-tree seeds and plants for windbreaks, shelterbelts and woodlands, and also in assistance to farmers in establishing, improving, and renewing such woodlands. The establishment of soil conservation districts under the Soil Conservation Act of 1935 encourages the farmer to use approved methods of preventing soil erosion on his farm. The Norris-Doxey Act of 1937 makes it possible to use his woodland more profitably in disposing of forest products. The Agricultural Adjustment Administration Acts enable the farmer to get special payments for improving and protecting his woodlot.

The U.S. Department of Agriculture, through the Forest Service, Extension Service, and Soil Conservation Service has carried on, in coöperation with states and farmers, five distinct farm forestry programs. In the fiscal year 1942 about $625,000 was available for this work.*

1. The production and distribution of forest planting stock to farmers was handled by the state foresters or other state forestry representatives in coöperation with the U.S. Forest Service. Farmers were able to obtain trees for reforestation of denuded forest land and eroded crop land at something less than the cost of production. In the calendar year 1941, nearly 98,000,000 trees were thus distributed in forty-two states. The federal contribution was $123,000 and the states' over $490,000.

2. Farm forestry extension work was carried on by the Federal

* Five distinct lines of work described by the Chief of Division of State Forestry, U. S. Forest Service, in a letter to the author dated July 16, 1942.

Extension Service coöperating with the State Extension Services in forty-two states, employing seventy Extension foresters. These Extension foresters carry on an effective farm forestry education program through the county agents and they also assist other state and federal agencies in the development of their farm forestry programs.

3. Farm forestry research projects are carried out in coöperation with the State Agricultural Colleges of eight states. These projects are designed to solve some of the many pressing problems associated with farm woodland management and most of them are economic in character, dealing with local utilization and marketing problems.

4. The Prairie States Forestry Project operates in the six prairie states of North Dakota, South Dakota, Nebraska, Kansas, Oklahoma, and Texas. The primary work (described in more detail in the chapter on federal forestry) is the production and procurement of planting stock and the planting and establishment of shelterbelts in coöperation with farmers and various agencies, to assist in improving and making more stable the physical, economic, and social welfare aspects of the region.

5. The purpose of the intensive demonstration projects established in coöperation with state agencies is to develop actual facts and figures on the value of well managed farm woodlands in the farm economy. Material assistance has been given to coöperating farmers in applying good forestry practices and in marketing farm woodland products. Thirty-eight states are coöperating and fifty-two such projects have been established.

Under the present Agricultural Conservation Program, soil-building allowances are available for about twenty-five different farm practices. Two of these have to do with woodland management. A payment of $7.50 per acre is made for reforesting open forest land with at least 1,000 trees per acre and a payment of $3 per acre is made for the improvement of the woodland by pruning or thinning under approved farm and wild-life management practices. In either case the plans must receive the approval of the forestry extension specialist or the county agricultural agent and local committee.

Experience has shown that the timber on a good woodland is often worth from three to four times the amount offered by timber buyers who buy by the acre or by entire tracts, without differentiating the species. If the advice of a forestry extension or marketing specialist is secured, markets can often be found for individual species used for pulpwood, piling, saw timber, railroad ties, etc., and for special high grade logs of such species as maple, black walnut, ash, or oak. This advice costs nothing and not only is the farmer financially better off, but his woods will be left in a productive condition looking to future harvests. Under the antiquated lump-sum method of disposal the chances are that his woodland would be stripped of every available stick of timber, both the ripe and over-mature and the vigorously growing young trees. The latter should have been left for the next cut.

The Farmer's Part in Supporting Forestry

It seems almost axiomatic that the farmer should be interested vitally in forestry. Woodlands occupy more acreage than any other crop on American farms and help support over 2,500,000 farm families. By working for forest conservation he is helping himself in the many ways just described. It is, of course, not claimed that forestry will cure all the farmer's economic ills. Complex problems like agriculture require complex solutions; the farm problem must be attacked from many angles. However, forest restoration can certainly help in a big way.

The farmer can aid the cause of forestry directly and indirectly by joining taxpayer associations to promote the wisest use of public funds and thus keep down local taxes. Farm organizations should introduce and back to the limit state bills for the purchase of forest land, for reforestation, and for forest fire protection. They should also back county, town, and local forests, especially as this means the taking of sub-marginal farm land out of production for reforestation purposes.

CHAPTER XI

FIRE, THE DESTROYER

1. MATCHES: Be sure your match is out. Break it in two before you throw it away.

2. TOBACCO: Be sure your pipe ashes and cigar or cigarette stubs are dead before throwing them away. Never throw them into brush, leaves, or needles.

3. MAKING A CAMP FIRE: Before building a camp fire, scrape away all inflammable material from a spot at least 5 ft. in diameter. Dig a hole in the center and in it build your fire. Keep your fire small. Never build it against trees or logs, or near brush.

4. BREAKING CAMP: Never break camp until your fire is out—dead out—cold.

5. HOW TO PUT OUT A CAMP FIRE: Stir the coals while soaking them with water. Turn charred sticks and drench both sides. Wet the ground around the fire. Be sure the last spark is dead.

6. BRUSH BURNING: Never burn slash or brush in windy weather or while there is the slightest danger that the fire will get away.

WHO HAS NOT READ THESE SIMPLE RULES for preventing fire in the woods? Yet records show that there was one forest fire every three minutes in 1937—over 185,000 of them. They burned $20,000,000 worth of timber on an area of 22,000,000 acres. The total economic losses ran into the hundreds of millions of dollars. And yet 1937 was a year of comparatively favorable conditions, for the number of fires was 18 per cent less and the area burned over about 50 per cent less than in 1936. We certainly stand indicted as a careless, thoughtless, and wasteful people. After over thirty years of unremitting and nation-wide effort to improve prevention and control techniques, our attitude still seems to be, "After us, the deluge."

The prevention and control of forest fires is a fundamental requisite for forestry, whether the principal harvest is timber, game, forage, recreation, or watershed protection. From the standpoint of forest conservation, fire is public enemy number one. The losses involved often far exceed the value of the timber burned. Forest fires destroy human life, wild life, and game. They not only destroy property but in so doing destroy income, dividends, and payrolls on account of their effect on wood-using and related industries. Most important of all, perhaps, is their destruction of the soil—our great national capital. The reduction of soil fertility is in itself a great national loss, but of far greater consequence is the erosion and flood menace that goes with it.

By what acid test can we etch these facts upon the public conscience? To many people, the relation of forests to their own welfare is vaguely abstract. To others it is cruelly concrete and real. If forest fires have destroyed their hunting; if muddy waters have spoiled their favorite trout stream; if floods have ruined their farm lands or drenched their homes with slime and mud; or if fires have destroyed their homes, livestock, or even their loved ones—no doubt they have been jolted out of their indifference and complacency. But how can we impress those who are careless with fire in the woods? Will it be more rules and persuasion, or will it be jail and fine?

What is the fire protection situation today in cold statistics? In the five-year period between 1926 and 1930 (more or less non-drouth years) there have been an average of 156,000 forest fires and about 41,000,000 acres burned over annually. This was on public and private lands throughout the United States. About 90 per cent of the fires were man-made. The total annual loss to industry, the public, labor, etc., probably amounted to several hundred million dollars a year. It is an arresting fact that 98 per cent of the area burned annually during these years was privately owned!

It has taken about thirty-five years to place approximately 130,000,000 acres of National Forests and 250,000,000 acres of private and state forests under protection. During 1932 our total fire protection bill was about $15,000,000 with the prospect that this will need to be materially increased in the future because there are still about 130,000,000 acres of private forests that are not receiving protection.

In European countries, in spite of a denser population, people for many generations have been taught to appreciate the intrinsic value of forests. A homogenous, intelligent, law-abiding population puts willful burning of the forests on a par with willfully setting fire to a man's house, barn, or crops. They consider such acts entirely out of the realm of possibility in a civilized state.

European experience shows that by proper police regulation, by silvicultural methods to reduce inflammability, and by proper management and organization it is possible to reduce the fire damage to an acceptable minimum. The proof is found in the forest fire statistics of the German forest administration. Over a span of twenty-five years the average acreage destroyed in 10,000,000 acres of Prussian and Bavarian forests rarely exceeded .005 per cent of the total area under state control, which means one acre in 20,000. On the National Forests of our own country, during the twenty-year period from 1910 to 1929, inclusive, the loss was .5 per cent of the gross area, or just one hundred times as much.

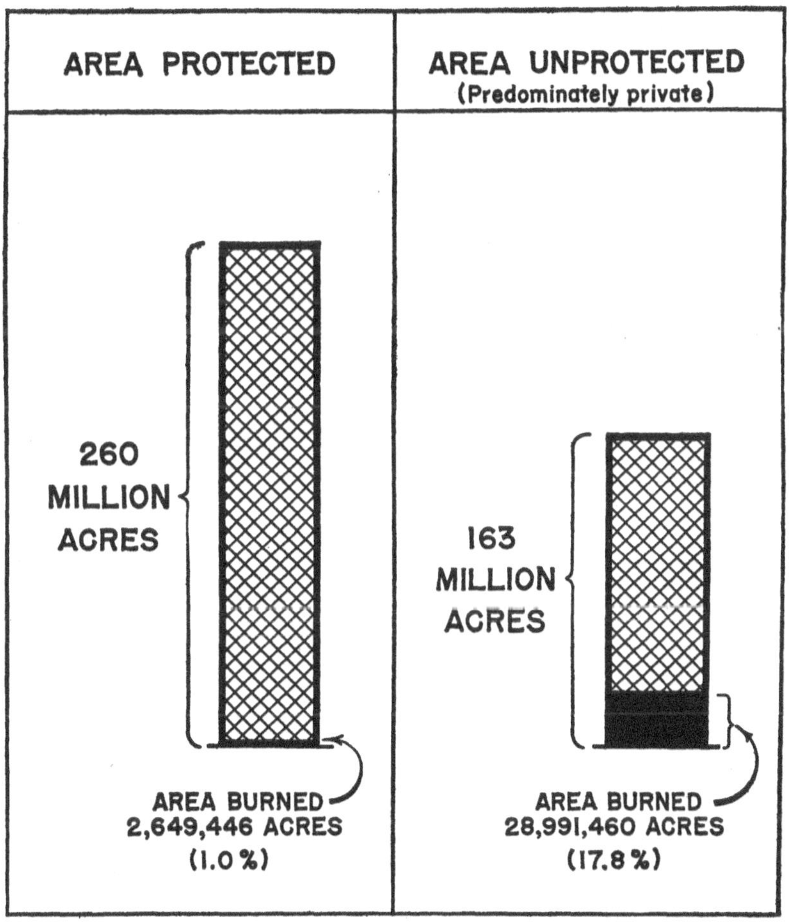

U. S. Forest Service Photo

PROTECTED VS. UNPROTECTED PRIVATE AND STATE LANDS

Average of Estimates 1935-1939

The greatest fire losses occur on unprotected lands

While these figures give an idea of the relative amount of fire damage under intensive European management as compared to that under extensive American management, it must be remembered that even this high rate of burn on our National Forests may approach an acceptable standard for management of our inaccessable, mountainous areas, where the factors for speedy control are often unfavorable. German forest-fire statistics, however, show what can be done.

To what extent have we succeeded in keeping forest fire losses down to an acceptable minimum? A certain degree of protection must be assured any timber area in order to safeguard the investment. *Complete* exclusion of forest fires, even if possible, would certainly be too expensive. Hence an objective in fire control has been set up by the U.S. Forest Service for each type of forest, based upon the percentage of the area that may burn over annually without interfering with the purposes of management. It is expressed as the area of "allowable burn," and it has been computed for each of the major forest types. It is interesting to know that this objective has been reached on the National Forests in some recent years, but that on private lands the actual burn has been eleven times greater, and in the southern states fourteen times greater, than the allowable burn under which forest productivity can be maintained. The unfortunate part is that we have reached somewhere near the goal of fire protection on only about 100,000,000 acres of National Forests, and have fallen far short of achieving the goal on some 400,000,000 acres of private forest land.

The beginning point of forest-fire protection is the realization that it is a tri-partite responsibility of the government, the public, and the private timberland owners. It must be predicated on the cooperative efforts of all the political, economic, scientific, and educational resources of these agencies, and the cost must be apportioned among them. All suggestions for a constructive program are made on the above premises.

The outstanding need in forest-fire protection is an intensive educational campaign to create a sympathetic public attitude for the purpose, principally, of reducing the man-made fires which constitute over 90 per cent of the total. Secondly, the solution of the private land protection problem calls for greatly increased appropriations to improve the fire-fighting technique, including better organization, communication, equipment, and supplies. Thirdly, we need more stringent laws about the use of fire in the woods, better law enforcement, and more severe punishment for carelessness and neglect, especially in the southern states. Finally, all agencies must realize that forest protection is going to be a problem of ever-increasing difficulty. A growing population, more leisure time, greater use of the woods, and larger areas of young inflammable growth and logging debris certainly will call for more vigilant prevention and more efficient control. There will probably be an increased use of fire breaks or fire lines to confine fire to small areas.

Historic Forest Fires

Forest fires belong in the same category as epidemics, famines, floods, earthquakes, droughts, and other human disasters. They have probably existed ever since there were forests. Geological evidences of this are presented in coal deposits, glacial drift, and charcoal peat bogs. Fire-scarred big trees of California show that great fires occurred in those forests in the years 245, 1441, 1580, and 1797. Fire scars on the Engelmann spruce show evidences of extensive forest fires in Colorado in the years 1676, 1707, 1722, 1753, and 1781. In the region south of Mt. Katahdin in Maine, white spruces record a forest fire covering 200 square miles in 1795. The assumption is supported by some evidence that prehistoric fires destroyed the forests that once covered parts of the great prairie region of the central states, some of the barrens of Canada, the extensive grassy parks of the Rocky Mountain states and the Appalachians, and the extensive treeless areas found below alpine summits.*

Many strange atmospheric phenomena are associated with great forest fires. These are variously known as "dark days," "red rains," "black rains," "black snow," and "dry fog." New England, which is in the path of forest-fire smoke borne by prevailing westerly winds, has experienced "dark days" in the years 1780, 1814, 1819, 1836, 1881, and 1894. "Black Friday" in New England occurred on May 19, 1780, and was caused by great forest fires in New York and Canada. In 1819 rain mixed with soot fell during a shower. In 1836 forest fires in Maine caused "dark days." In 1881 fires in Michigan caused "dark days" in most of the northeastern states. The great Hinckley fire in Minnesota darkened New England in September, 1894.†

Such heavy palls of smoke interrupt business, interfere with navigation, compel the use of artificial light, and injure health. Breathing is made more difficult and the caustic properties of the smoke and soot often cause a burning sensation in the eyes and air passages.

Statistics fail to reveal the appalling losses in timber, homes, wild

* F. G. Plummer, *Forest Fires* (U. S. Forest Service Bull. 117, 1912), pp. 7-8.

† J. D. Guthrie, *Great Forest Fires of America* (U. S. Forest Service, 1936), p. 3.

life, and human lives which this country has suffered in the past. If we could compile data covering only the last hundred years they would be beyond human comprehension. Plummer in 1912 made an attempt to gather the known data on forest fires in his now-famous Bulletin 117. This records several thousands of the more conspicuous fires that had occurred from about 1825 to 1911. Since, however, no real records were kept in the western states until about 1905 and these cover only government lands, the entire history of forest fires is fragmentary to say the least. It has often been stated that many times as much timber has been burned than has been cut since white men settled the continent. A brief chronological résumé of some of our greatest forest fires will hit only the "high spots," so to speak, but will be sufficient to show that there has been colossal carelessness and waste.*

It is of course useless to read over a lot of figures on losses if the reader does not realize what they mean. What is a million acres? How much timber could there be on it? How many houses could the timber build if it did not go up in smoke? Do fire losses consist merely of the loss of stumpage and so much property?

It takes an average of about 5,000 board feet of lumber to build a six-room house. An acre of western forest may have on it from 10,000 to upwards of 200,000 board feet. A million acres is an area 10 miles wide and 156 miles long. A fire destroying a million acres of timber would destroy the lumber capable of building 2,000,000 homes, assuming that the average timbered acre contains only 10,000 feet.

The losses given in terms of stumpage value are misleading, because that value is only a few dollars per thousand feet. The actual loss is really at least four to five times that amount, and in a big Oregon fire was computed to be about seventeen times the stumpage value. If the lumber had been cut and put on the market as manufactured articles, there would have to be added the costs due to investments in mills and wood-working factories, the wages paid to labor, and various other manufacturing and transportation costs. Such industries also pay taxes to support roads, schools, and other improvements and pay dividends to investors. The financial ramifications are wide and numerous, and are only too seldom considered in forest-fire statistics.

In 1825 there occurred the famous Miramichi fire in Maine and New Brunswick, sweeping over 3,000,000 acres and taking 160 human lives. In 1846 fire swept through 450,000 acres of some of the densest Douglas fir forests on the Pacific Coast. This is recorded as the

* Taken from J. D. Guthrie, *Great Forest Fires of America*, pp. 4-10.

Yaquina fire. In 1849 Maine was again visited by conflagrations when over 4,000,000,000 feet of timber were burned on 1,000,000 acres. In 1853 the Nestucca fire in Oregon covered 320,000 acres, and in the same year the Pontiac fire in Quebec burned over 1,600,000 acres. In 1865 Oregon again lost dense stands of timber over an area of 1,000,000 acres in the famous Silverton fire. Two fires in Oregon in 1868—the Coos and the St. Helens—covered 600,000 acres. For over a month western Oregon and Washington were enveloped in a thick pea-soup smoke, giving rise to so-called "dark days."

The most calamitous fire in our history, as far as records show, was the famous Peshtigo fire in Wisconsin in 1871 which burned over 1,280,000 acres, sweeping many homes and towns and killing over 1,500 persons. In the same year fires in Michigan covered 2,000,000 acres. The famous Bighorn fire in Wyoming in 1876 burned over 500,000 acres. In 1879 New York State lost over 3,000,000,000 feet of timber over 1,350,000 acres. One of Michigan's worst conflagrations occurred in 1881, when one million acres were burned with a loss of $2,000,000 and 138 human beings. In 1894 the Phillips fire in Wisconsin burned 100,000 acres and killed 300 persons. In the same year the great Hinckley fire in Minnesota wiped out twelve towns, 160,000 acres of forest, and over 400 people. In 1896 West Virginia lost about 3,000,000,000 feet of timber in a disastrous fire. In 1902 a gigantic fire in Washington and Oregon destroyed 600,000 acres and $12,000,000 worth of property, while the next year saw one of New York's worst fires in which 450,000 acres were devastated, mostly in the Adirondack Mountains.

The summer of 1908 was a disastrous year in West Virginia, Wisconsin, Michigan and a number of other states. In these three states alone over $10,000,000 worth of standing timber was burned. In 1910 Minnesota, Montana, Idaho, Washington, and Oregon suffered tremendous losses. Over $10,000,000 worth of timber went up in smoke and over 100 lives were lost. From 1906 to 1910 California lost over 2,500,000 acres of timber by fires.

An exceedingly dry year in Minnesota caused the Cloquet fire with numerous others in 1918. Cloquet, a busy sawmill town of 12,000 people, was wiped out and timber and property valued at $30,000,000 was lost, to say nothing of some 400 lives. At one stage of the holocaust the big city of Duluth itself was threatened. In 1932 there was a disastrous 220,000-acre fire in the Matilja Canyon on the Santa Barbara National Forest in California. It was fought by 2,500 men. Five portable radios and two airplanes were used, and over 300 miles of fire lines were dug. The fire did enormous damage to the watersheds of eight towns and cities and to important irrigated lands. The loss of life among deer, quail, and other wild life was very heavy. Most of the burned area was brush of little commercial value, but of great value for protecting watersheds from erosion and floods.

One of the most serious of recent fires was the big Wilson River fire of August, 1933, in Oregon, started by friction from a steel cable passing around a dry stump. Over 260,000 acres of the finest virgin timber in the state was burned, valued on the stump at over $20,000,000. The total loss to the industry, the public, and to labor was computed to be in excess of $350,000,000. It is interesting to note that in this fire alone an amount of timber was destroyed equal in amount to the entire timber cut of the United States in the year 1932, namely, 12,000,000,000 board feet.

Forest Fire Losses

It has already been pointed out that aside from the toll of human lives, the greatest apparent losses due to forest fires are loss of stumpage, the burning of homes and business properties, the destruction of wild-life and recreational values, the irreparable damage to watersheds, and the widespread economic losses suffered by industry, labor, and the general public. In all these losses we are reminded that raw materials used by many industries are destroyed. These industries would have used this burned-up wealth to put people to work, fill pay envelopes, pay taxes, interest, and dividends, and finally to make people happy and healthy.

Stupendous as these losses are, they are only a part of the sad picture. Let us look further into just one item of loss above, namely the damage done to watersheds when fire removes the protective cover. Damage to such watersheds, it must be emphasized again and again, increases erosion and floods. Such damage has a far-reaching effect on industries which depend upon a regular water supply such as agriculture, navigation, irrigation, hydro-electric power, and domestic water supplies. The people of southern California know this only too well. Water is the life-blood of this region. Any vegetative cover—forest or brush—which conserves and regulates the water of that semi-arid region must be considered as essential as the heart is to the human body.

Finally, we come to the most far-reaching effect of all, namely, the destruction of the productivity of the forest soil. Many of us have noted how slowly forest growth comes back on an area swept bare by fire. It may take years for valuable forests to get started. Careful observations have shown that it may take thirty to forty years for the original species to reëstablish themselves. Many heathlands, moors, prairies, oak barrens, "Karst" regions, sandy barrens, brushlands, and other economic wastes the world over are due to changes in soil fertility, in many cases caused by repeated forest fires. The humid prairies of the Middle West, the oak barrens of Michigan, and certain brushfields in California are mainly due to forest fires.

Fire often destroys completely the favorable physical, chemical, and biological properties of the soil.* Forest soil in its best condition is loose and flaky, water-absorbent, rich in chemical compounds taken in by the roots for growth, and teeming with soil bacteria and soil animal life which help to make these food materials available. A burned soil is sterile, with little water-holding properties, and lacking in organic matter which is the material on which microscopic plants and animals feed. Nitrogen, the most essential element for life and growth, is lost in the smoke of the fire and is replaced only after decades of accumulation following a succession of vegetation.

Furthermore, if the destructive forest fire happens to be on a slope, rains will soon carry the soil away, exposing bedrock or hard pan, neither of which will support good plant growth. Government observations show that forest fires may increase run-off as much as eighty-five times, and the erosion that follows may be increased in much greater proportion. Aside from the loss of valuable topsoil, this and other debris may cover fertile farm lands in the valley below—and here is where forestry and agriculture are inseparably linked.

History shows that civilization depends on soil fertility. This applies to both agricultural and forest soils, since the former give us our foods and the latter wood and water. All three of these—food, wood, and water—are essential to human civilization. Once the topsoil is stripped of that life-giving humus and the denuded slopes permit a swift run-off of water, agriculture and other valley occupations become precarious, good farms become border-line farms, landowners become discouraged, tenancy increases, and the net result is social and economic degeneration. The history of nations shows that agricultural impoverishment over large areas has been an important contributing factor in the downfall of ancient civilizations. China, Asia Minor, India, Greece, and Rome furnish us with horrible proof of the old saying that "When the soil goes, men go." †

* *National Plan for American Forestry* (Sen. Doc. 12, 73rd Cong., 1st Sess., Washington, D.C., 1933), pp. 310-320.

† Lois Olson, *Erosion, A Heritage from the Past* (U.S. Dept. Agr., 1940), pp. 1-12; W. C. Lowdermilk, "The Eleventh Commandment," *American Forests*, unnumbered reprint, January, 1940.

Causes of Forest Fires

Forest fires occur in all the forest regions and on all classes of lands, regardless of ownership. There is a great difference, however, in the intensity of the fire problem as between regions. Statistics of the average annual burn on National Forests, state forests, and private lands for the period 1933-37 show over 36,000,000 acres burned over by more than 172,000 fires. About 63 per cent of the total forest and potential forest land requiring protection is actually receiving some form of protection. This ratio is lowest in the south and central states and highest in New England and the Rocky Mountain states. About 130,000,000 acres of forest land still receives no form of protection.

The most arresting fact concerning the causes of forest fires is that over 90 per cent of them are man-caused. The following table shows the percentage of fires for each of the major causes, covering protected areas on national, state, and private lands for the period 1933-37:

Cause	Per cent of total
Lightning	8.7
Railroads	4.3
Campers	6.3
Smokers	24.4
Debris burning	13.7
Incendiary	24.7
Lumbering	1.8
Miscellaneous	8.9
Unknown	7.2
Total	100.0
Man-caused	91.3

The greatest lightning areas are the Pacific Coast and Rocky Mountain states. In the southern Rocky Mountains, lightning caused over 60 per cent of the fires. In the eastern states, fires due to this cause are negligible. Railroads are an important cause of forest fires in the New England, middle Atlantic, and Lake states. Campers seem to be about equally negligent in all regions except New England, where only about 2 per cent of the fires are due to this cause. Smokers cause a larger percentage of fires in the New England, middle Atlantic, Lake, and Pacific Coast states than in other regions. Incendiary fires are numerous in the southern and central states. In the Rocky Mountain states, man-caused fires are about 36 per cent of the total; in the Pacific states this figure jumps to over 76 per cent; while in each of the eastern regions the percentages of man-caused fires is over 98 per cent.

Fire Hazard

The geographic distribution of burned-over areas shows that 90 per cent of the total area is in the southern states, where less than 25 per cent of the forest land receives protection. The central and Pacific states rank next in point of area burned over, and the southern Rocky Mountain and New England states have the best record from this point of view.

Closely related to the fire hazard in various regions is the length of the fire season based, of course, upon the length of time each season during which the forest remains dry and inflammable. While this varies somewhat from year to year, it is very evident that some regions have a much longer season for forest fires than others. In many parts of the West, July and August are the months of greatest danger. In the East there are usually spring and fall fire seasons in addition to occasional summer danger. In the South the fire season often starts in February or March and may last until late summer.

Behavior of Forest Fires

The amount of damage done by a fire cannot usually be judged by its area alone, because it depends upon where and what it burns. Ground fires burn in the humus and peat of our northeastern forests. When swamps dry out, the peat or sphagnum may catch fire and burn slowly, in many cases underground, for a long time. Such fires may not do much damage, but their chief danger lies in escaping unnoticed and later developing into disastrous surface or crown fires. Surface fires burn close to the surface of the ground, consuming needles, young growth, dry grass, brush, and down timber. They often do great hidden damage, because they deplete the fertility of the soil, injure growing timber, and cause infection courts for insects and fungi. Crown fires are most disastrous of all, since they travel in the crowns of evergreen trees, which are highly inflammable on account of their resin content. The Pacific Coast, the Rocky Mountains, Lake states, and northeastern states suffer greatly from this type of fire.

The severity of a forest fire and its tendency to spread depends upon a number of factors, the chief of which is the inflammability of the stand. Hardwood or deciduous stands are less dangerous than the coniferous. A long-continued dry spell, low atmospheric humidity, high winds, and steep topography all aid in spreading the destruction. Narrow valleys and canyons become huge fireplaces, and the fire creates its own draft as the hot air rises. A horse at full speed sometimes cannot outrun the fire demon.

The Forest Fire Problem

The problem of forest fires is much easier to state than it is to solve. Briefly it is this: to build up, by means of every educational device known, moral, legal, and financial support to prevent and suppress forest fires.

Obviously, the first step is to acquaint the public with the facts. This entails an educational program far in excess of anything before attempted. As this subject properly falls under a later chapter, only a few suggestions will be made here. This education must be motivated. Whether a man builds a house, goes fishing, goes hunting, runs a business, runs a farm, goes camping, runs an automobile, buys a summer home—or whatever he does, the forester must show how the elimination of forest fires comes into the picture. Furthermore, I should like to see laws passed to make the teaching of forest conservation compulsory in the public schools of every state in the union. Georgia, Louisiana, Mississippi, North Carolina, and Tennessee have such laws now. Once the facts are generally known (and this volume attempts to give them), the moral, legal and financial approaches to the problem will be made that much easier.

Approaching the problem from the moral angle is probably most difficult of all. Some individuals need to be taught that democratic freedom must be tempered with restraint, and that coöperation with constituted authority is necessary to the continued enjoyment of the rights and privileges which are denied citizens under other forms of government. If such coöperation is not forthcoming, the only alternative is restriction or denial of such privileges, and, in extreme cases, fines and imprisonment.

Fire Prevention

Forest-fire prevention must be predicated on four broad major premises. First, there must be inaugurated an educational campaign to stimulate a sympathetic public attitude towards the whole forestry problem. Being essentially a problem "of the people, by the people, and for the people," the general public must understand what federal, state, and local authorities are trying to do. Second, the enactment of organic laws by the states must make prevention and control a state responsibility. Organic laws have been functioning on federal lands for the past thirty-five years, but these constitute only about one-fifth of the forest land needing protection. Third, for those who cannot become forest-fire conscious as fast as they should for the good of society, we must enact further stringent laws—federal, state, and local—to punish carelessness and incendiarism. Lastly, any government unit undertaking this work must be assured continuous financial support to maintain an adequate and efficient protection organization.

A considerable number of laws, regulations, and practices have developed on the National and State Forests as well as on private lands to prevent forest fires. Every user of the forest should be familiar with them so he can coöperate to make them effective:

1. Campers are restricted in the use of fire in the woods. Permits are sometimes required to build camp fires, if they are not built on specially designated areas. Campers are often required to register with the official in charge. Usually camp fires are restricted to specially prepared and designated areas prepared in advance by the forest officials.

2. Smoking in the woods may be restricted or prohibited in times of great fire hazard.

3. There is usually frequent contact and inspection by patrolmen, guards, and rangers to enforce the laws and regulations.

4. Campers are often required to carry fire-fighting tools in their cars, such tools being the ordinary ones used in camping as well.

5. There may be a complete closing of forest areas to all forms of tourist traffic in times of great danger. This may also include hunters and fishermen.

6. Often the attempt is made to keep the traveling public in touch with the forest fire situation by various educational devices such as newspapers, radio, folders, road signs, etc.

7. In coöperation with the local Weather Bureau offices, weather predictions warning of unusual fire hazard are disseminated in the same way.

8. It is often necessary to control fire used in logging operations, and in extreme cases stop operations altogether.

9. Railroads may be required to have equipment available to help check fires such as spark arrestors in locomotives, water tanks, pumps, and hose and fire-fighting tools.

10. In areas suitable to their use, firebreaks, lanes, plowed furrows, etc., are constructed and maintained in advance of their need.

11. Much fire hazard can be eliminated by the disposal of brush and other logging debris which forest fires feed on.

12. This is especially necessary along much-traveled roads. Caches

of tools and equipment, fire fighting machinery, and food are stored at strategic points for quick distribution to needed points.

13. Putting out a fire in its incipient stages is in one sense prevention, and for this purpose it is necessary to maintain means of detection, communication for reporting, and transportation.

14. There must be continuous research in prevention methods with special attention on the causes of forest fires and the improvement of methods based upon experience.

15. There must be prepared detailed fire control plans for each ranger district or other unit which can be put into immediate operation at a moment's notice. Detailed plans for coöperation with private, state, and federal agencies and local communities in all phases of fire protection must be included in all fire control plans. The whole scheme must work with the precision of a military machine.

After a forest fire starts the problem is to get men, tools, and other equipment to it in the shortest possible time. The fire control problem therefore resolves itself into sub-problems of detection, communication, transportation, availability of tools and equipment, and actual suppression.

Detection

The detection of forest fires is accomplished on National Forests, National Parks, State Forests, and other organized protection areas by lookout men stationed on high peaks or on other elevations which give an unobstructed view of the surrounding country. There are over 3,000 of these lookouts on the National Forests alone.

Men or women assigned to this duty live in towers, cabins and other structures, some of which are unusual to say the least. One early lookout station known to the author was perched about two-thirds of the way up a giant cedar tree, with a crude ladder leading to it. It was nicknamed "the strawberry box." Many of these lookouts have interesting histories and have been erected under great difficulties. The one first erected on Mt. Lassen, California, in 1913, was built in a carpenter shop in a town seventy-five miles away, transported to the foot of the peak, carried piece by piece on men's backs for the last two miles, and re-assembled on the highest point at an elevation of 10,500 feet. The very next year this structure, erected so laboriously, was destroyed by volcanic action.

Such lookouts usually are equipped with maps, binoculars, an Osborne fire-finder, and a telephone connected with the nearest ranger station or other habitation. If the lookout man also lives in the lookout he is provided with a stove, folding table, bed, chairs, and other simple conveniences.

Fires are also detected by patrols of fire guards and other forest officers who move by automobile, on foot, or on horseback along forest trails and roads. For detecting and reporting forest fires, the coöperation of tourists, hunters, fishermen, logging companies, cattle and sheep men, and other forest users is secured as far as possible. Finally airplanes, both land and water, are increasingly used in Canada and parts of our West to locate fires in inaccessible level areas where lookouts are lacking.

Communication

The report of the location, nature, and extent of the discovered fire is usually phoned to the ranger in whose district the fire is burning, or to the forest headquarters where the supervisor will receive it, or to some other forest officer who is temporarily stationed in the vicinity. In areas where there are no telephone systems, heliographs using the Morse signal code are sometimes used. Help from outside areas may be summoned by telegraph, and the portable radio has been of great service in reporting the progress of a fire and thus assuring a coördinated attack. Communication is well planned in advance, and automatically the fire suppression system goes into action in high gear as soon as the fire is located.

Transportation

The next step is the transportation of men and equipment to the scene of the fire. This is usually accomplished by passenger auto or truck, at least to the point in the highway nearest the fire. In remote regions where only trails are available, horses and pack animals are employed. Men from outside are often hauled in on logging trains. Airplanes have been used to study the behavior of fires and to drop food and fire fighters in wilderness regions. These "smoke jumpers" are dropped by parachutes in inaccessible regions and are furnished with necessary equipment including a short-wave radio for two-way communication. Motorcycles and railroad speeders have been used with much effectiveness in certain regions.

Man Power

Local rangers and guards are considered the first line of defense in this battle against forest fires. If the fire warrants it, men from local communities and industries may be called in, and as a last resort labor from the large cities, or even soldiers from the United States

Army. During the years 1933-1942 the Civilian Conservation Corps boys located in camps throughout our vast forest regions were the principal source of man power.

Equipment

Fire-fighting equipment must, of course, be readily available at ranger stations, roadside caches, camp sites, summer resorts, lumber camps, and other convenient places. The most usual tools are axes, shovels, mattocks, grub-hoes, water buckets, hand water-pumps, bucking saws, and even power pumps and hose. If the fire is large the central supply depot for the administrative district dispatches tools and equipment in larger quantities by rail or by truck. Such equipment, in addition to the tools mentioned, may consist of food supplies, motor plows, fire trucks and pumps, tractors, tents, bedding, cooking outfits, riding and packing equipment, and other equipment especially designed to meet the conditions.

Every small fire must be considered an incipient conflagration. Local officials must be prepared either to suppress a small fire, face a broadside of fifty lightning fires started at once, or to cope with a conflagration requiring a thousand men over a period of several weeks.

Suppression

Actual suppression of a small fire is begun with tools like shovels, mattocks, grub-hoes, axes, and saws. The fire is smothered with dirt or sand, and inflammable materials such as fallen logs, branches, dry leaves, and brush are removed around the fire. Under certain conditions a direct attack can be made with water if it is available. Clearing fire lines is the most effective means of stopping a widespread surface fire. All inflammable debris is removed down to mineral soil and thus the fire is completely surrounded by a cleared line. Surface grass fires are sometimes smothered with wet burlap bags or blankets. Large fires, especially if burning in the crowns of the trees, are sometimes controlled by back-firing from a road, railroad, fire line or firebreak. Thus fire is used to fight fire, but this tool should be employed only by experts, for it is a dangerous procedure. Crown fires are the most difficult to stop. Often when fanned by a high wind little can be done directly to check the flames. These fires are more common on the Pacific Coast and in the northern Rocky Mountain states where pure stands of evergreen forests are a great fire hazard.

Progress in Fire Protection

The technical knowledge and skill we possess today with reference to all the major phases of the forest fire problem have accumulated since 1910, the year of the great forest fires in the Pacific Northwest. When the Forest Service took over the administration of the National Forests in 1905 there was little technical knowledge extant in America for protection against fire, or adequate tools, equipment, and personnel for the job. The body of knowledge and experience gained during the last thirty years in this country is unique in forest protection history, both on account of the great diversity of its application and on account of the originality of its conception and execution. No other forest country in the world could furnish us with the ways and means for attacking the particular problems determined by our peculiar combination of topographic and climatic conditions.

During the last thirty years several factors have made fire protection more difficult. There has been a much greater use of the forests by the public and an increased incendiarism that has been aggravated at times by large-scale unemployment. There has been a constant increase in cut-over areas with their attendant hazards of dangerous slash. Added to these, there has been from 1916 to the present time a continued and cumulative deficiency of precipitation over many of the dangerous regions of the West.

On the other hand, there has been a great increase in the total forest area protected and also a much better reporting of fires that have occurred. The result is that we now have a much more complete forest-fire record to work with. The records show that there is a general downward trend in both the size of the average fire and the annual forest area burned over.

CHAPTER XII

INSECT ENEMIES

There are between 300 and 400 species of forest-tree insects in this country, and practically no important timber tree is immune from their attack. The damage done is variously estimated at from $100,000,000 to $150,000,000 a year. Two general kinds of outbreak occur. The greatest losses are not caused by the conspicuous local onslaughts that attract wide attention, but by the sustained annual loss of scattered merchantable trees over large areas. Local infestations often kill a large percentage of the trees of one or more species in a small area, while the scattered infestations that kill a tree here and there, though they often escape unnoticed, add up to a tremendous total.

Some Conspicuous Outbreaks

Forest-tree insects may attack any part of a living tree—root, stem, leaves, flowers, fruits, seeds—whether it be in the seedling, immature, or mature stage. They also attack logs in the woods, seasoned lumber, and the finished wood products. History furnishes us with numerous examples of widespread insect depredations. Between 1882 and 1885 the larch saw-fly defoliated our eastern larch or tamarack in New England, Michigan, and Canada. The southern pine beetle first became conspicuous about 1891, and since that time has taken an enormous toll. The Black Hills beetle is said to have killed about $2,500,000 worth of pines in the forests of the Black Hills of South Dakota between 1898 and 1908. In New York and New England the spruce beetle has done sporadic damage for over a hundred years, often killing eastern spruces over wide areas. Similar damage has been done to the Engelmann spruce forests of the Rocky Mountains.

All these insects and many others are still with us. It has been estimated that the western pine beetle destroys annually over a billion feet of western yellow pine in California alone. The mountain pine beetle has made serious inroads on the lodgepole pine in the

West, the author having supervised an estimate of damage done by the beetle covering 100,000 acres in northern California. The spruce budworm has destroyed millions of cords of spruce and balsam in the East in the last twenty-five years. The gypsy and browntail moths have spread rapidly through New England defoliating deciduous trees, and over $7,000,000 has already been spent in their control. Termites have become a serious menace to wooden foundations in the South and in California. The white pine weevil, working on the shoots and leaders of white pine, affects the growth and form of this important timber tree. Other more or less important insects are the locust borer, hickory bark beetle, the pine butterfly, the hemlock looper, and the numerous insects which damage felled logs, sawed lumber, and finished wood products.

The Extent of Insect Damage

The nature and extent of insect damage depends, of course, upon the part of the tree attacked. A thorough knowledge of these various methods of attack is important in most phases of prevention, detection, and control. Without this knowledge foresters are merely groping in the dark. The life history of each important insect should also become common knowledge to all intelligent forest users. A number of forest tree infestations have been reported to state and federal authorities by hunters and fishermen.

Insects devour the leaves of trees and may entirely defoliate them, thereby destroying the very organs that make the food. They may suck the sap from the tender young leaves and shoots, thus lowering the vitality of the tree and decreasing the current year's growth. Other insects may attack the flowers, fruits, and seeds of the tree, thus affecting the power of reproduction. Probably the greatest damage is done by the bark beetles which infest the cambium and outer sapwood and interfere with the conduction of water and food. These may kill the tree and thus entail direct financial loss.

The damage done by insects to living trees is not entirely confined to the loss of timber. Damage often causes the loss of annual growth, which, if it affects large areas, may be considerable. The infestation of trees by one class of insects often invites other insects. Leaf-eating insects are often followed by bark beetles and these by wood-burrowing insects. Thus the trained entomologist differentiates between insects that cause the primary injury and those responsible for the secondary attacks. Insects have a way of infesting trees whose vitality has been impaired either by other insects, by fire, or by fungi. Hence we see the necessity of keeping the forest in a healthful condition and of preventing these agencies from impairing the vitality of the trees and thereby inviting the work of destructive insects.

Recognizing Insect Attack

The question arises as to how an insect infestation can be spotted by a forest owner, a ranger, or a layman. This is rather difficult to answer because of the many kinds of damage done. Let us take as an illustration the work of a typical bark beetle. Reddish brown masses of pitch and sawdust on the bark of a tree are easily recognized as the work of this insect. A pine or a spruce tree with all its needles turned yellow is also a warning of bark beetles. In this case the tree was probably attacked during the previous summer. If the needles are red, the tree was probably attacked more than a year ago, since trees attacked in the spring of one year usually do not show the effects until the following summer. These two stages are known to the entomologist as the "yellow-top" and the "red-top" stages. After a number of years the dead trees stand out very conspicuously as leafless gray or black snags, when they are said to be in the "black-top" stage.

Now as to pitch tubes and sawdust on the bark. These are small reddish-brown (later becoming grayish-white) masses of pitch and sawdust which are expelled from the small cylindrical entrance made by the adult beetle where it bores through the bark to begin its egg tunnel. Each tube represents the entrance of one or more of these beetles. As soon as the beetle has made its entrance through the bark, it starts to work up through the live bark, cambium, and outer layers of sapwood, forming a tunnel, but little larger than the diameter of the beetle, known as the "egg gallery." Such egg galleries vary in shape from straight to winding and in length from ten to forty inches. As a rule male and female beetles work together in one gallery and the eggs are deposited along the side of the gallery, often in little pockets. When the tunneling and egg-laying process of the adult beetles is completed, their life work is over and they are usually found dead at the upper end of the gallery. The larvae hatch and begin their work by burrowing across the cambium at right angles to the egg galleries. The complete girdling of the cambium region is not accomplished until the larvae have completed their work, and the numerous larval galleries, by joining one another, form a complete gallery around the tree, thus cutting off the food and water supply, preventing water from moving up to the leaves and food from going down to the roots. The tree usually dies the following spring. As soon as the larvae have completed their development, they pupate and later become adults which swarm forth in huge numbers to attack new trees. Needless to say, if such trees could be spotted, cut down, and burned before the new beetles emerge the whole new brood could be destroyed.*

Prevention and Control

The prevention and control of insect damage is a highly complex problem. Although various agencies have been working for over thirty years on these problems, only within the last fifteen years has sufficient money been spent to make even a slight impression. The general condition of our forests from a sanitary point of view is so unbelievably crude that the forester is puzzled to know where to start. Fortunately the insect problem can be attacked from a score of angles both direct and indirect, and it is being attacked in that manner today as far as the appropriations will permit. The work is difficult, slow, and at times discouraging. It takes a great deal of courage and perseverance to carry on.

Let us first consider control methods which have to do with general forest policy and which are broadly applicable to the country as a whole. As already pointed out, there must be increased appropriations for all forestry agencies. Adequate research is needed on methods of prevention, detection, and control. A thoroughly detailed and adequate system of detection in the field is essential to secure early recognition of insect attacks, so that control measures can be speedily instituted. Further research, which should be coördinated by all agencies, is needed to complete the determination of facts concerning the life histories of American forest insects. Such knowledge should be disseminated to all forest officers and also, in a less technical way, to all forest users. Federal quarantine must help to prevent the introduction of any more foreign pests, and must prevent the spread of insects from one state to another as far as possible. Few realize the difficulty of this problem since the introduction of the automobile and the airplane. Finally, an intensive campaign should be waged to increase the use of methods for preserving wood, in order to control such wood-boring insects as marine borers and termites.

Measures of preventive control are found in good silvicultural practice. First of these, of course, is fire prevention, since fire weakens trees and opens the way for infection. Proper forest sanitation methods must be instituted, since insects attack weak, dead, and dying

* For a fuller treatment of insect enemies, see A. D. Hopkins, *Bulletin 48*, Bureau of Entomology, 1904, and the same author's "Insect Enemies of Forest Reproduction," *Agricultural Yearbook*, 1905. Also, S. A. Graham, *Principles of Forest Entomology*, New York, 1939, and S. W. Frost, *General Entomology*, New York, 1942.

trees. The removal of such trees would be an important step. Where severe depredations have occurred the timber should be sold, if possible, and the bark and slash burned. It is possible to reduce the proportion of species subject to insect attack by cutting and selling the more susceptible species. Forest plantations of such species should be discouraged. Mixed forests are less liable to destructive depredations than pure forests. In general large areas of pure stands can be avoided in certain areas. The handling of logs, semi-finished, and finished products must be improved so as to keep insect damage down to a minimum.

Methods of direct control are expensive and often impossible of adequate application. At present active eradication is going on for the gypsy moth, the elm-leaf beetle, and some others in the Northeast. The egg, larva, pupa, or adult stages of insects can be collected and destroyed. The importation, in the case of foreign pests, of their natural parasitic and other enemies seems to be the best method of controlling these insects. Thus the Calosoma beetle has been imported to combat the gypsy moth. For native pests, the protection and encouragement of their natural enemies, especially birds, is highly desirable. Insecticides are of only limited use as far as we now know, principally on account of the cost. Spraying and dusting from airplanes may hold out some hope for success if such operations could be scientifically checked for results. Deliberately weakening the vitality of certain trees by girdling so that they will attract broods of insects has been done. The trees are later burned, broods and all. This "trap-tree" method is of limited application.

We may attempt a rough summary of the insect problem in the following key words: prevention, sanitation, quarantine, detection, suppression, coöperation, research. The entire subject is so technical in character and so nation-wide in scope that nothing short of full and complete coöperation of all agencies can hope to be effective in the seven lines of endeavor indicated above. *Sanitation* should be one of the first duties of every forest land owner. The U.S. Bureau of Plant Quarantine is *preventing* the importation of more harmful foreign insects at the ports of entry. *Detection* and *suppression* are carried on by federal, state, and private agencies in coöperation with the U.S. Bureau of Entomology. Suppression is largely the task of land-owning agencies, private, state, and federal. Federal insect control work is being carried on by the Forest Service, the National Park Service, and the Bureau of Indian Affairs. A *coöperative* agreement between the federal government and the states similar to that embraced in the Clarke-McNary Act, whereby the federal government matches the state appropriations for fire control, is needed for insect control. Federal aid is now given to states on specific projects, and state aid has been extended to private owners, especially in the gypsy moth and elm-leaf beetle control work being done in some of the northeastern states. *Research* is being carried forward by federal, state, and private agencies, but principally by the first named. A wider scope of work is needed in entomological research at the Forest Service Experiment Stations, at the State Agricultural Experiment Stations, and at the State Universities.

CHAPTER XIII

TREE DISEASES AND FOREST SANITATION

THE PROBLEM of forest-tree diseases is in many respects similar to that of forest insects just described. The same tremendous, incalculable losses by a more or less hidden enemy are noted. All parts of the tree are attacked, as well as cut material in the woods and the finished products. Losses due to forest insects and diseases in the aggregate exceed those from fire.* The methods of prevention, detection, and control are very similar, and complete information of the life histories of the causative organisms is notably lacking. The suppression and control of forest-tree diseases also aim at better forest sanitation, the suppression of forest fires, adequate detection facilities, quarantine, better coöperation among timberland owners, more research, and more adequate appropriations.

Two Devastating Tree Diseases

One of the worst diseases to menace stands of American timber is the chestnut bark disease, discovered about 1904 near New York City by my old friend and teacher, John Mickleborough. It was probably introduced from China. At the present writing it has destroyed most of the commercial stands of chestnut in New England, New York, New Jersey, Pennsylvania, and the southern Appalachians, and is threatening the few remaining stands. This thoroughly devastating disease causes the death of the chestnut by destroying the cambium and inner bark. Spores attack a tree through a break or wound, at which point they germinate and produce a fungus growth which finally kills the tree by girdling. The only part of the fungus visible to the unaided eye is the yellowish or orange-colored pustules or fruiting bodies which protrude from the bark. The fungus is rapidly disseminated by spores carried by the wind and by birds, and it attacks trees of any age. No method

* *Forest Lands of the U. S.*, House Report 323, 1941, p. 2010.

of control has yet been discovered. Immune varieties are being developed by crossing the American with Japanese varieties.

No doubt the most menacing forest-tree disease today is the white pine blister rust imported into this country on nursery stock from Germany about 1909. It is now threatening our five-needle pines—white pine, western white pine, and sugar pine—whose commercial stands are valued at over $500,000,000. It is now known to exist over most of New England, New York, New Jersey, and Pennsylvania, the Lake states, and Canada. Introduced also into British Columbia in 1910 it has spread rapidly into several western states. It is interesting to know that the cultivation of American white pine in certain parts of Germany, Holland, and Russia has been abandoned due to the work of this fungus.

The white pine blister rust, like the wheat rust and similar fungi, requires two hosts to complete its life cycle. It lives alternately on various species of native and cultivated currants and gooseberries and the five-needle pine. Removal of currant and gooseberry bushes in the vicinity of white pine stands has proved an effective means of control.

Infection by this fungus can be noted by the swelling of the young stems of the white pine and the subsequent development of orange-yellow pustules or fruiting bodies which discharge their countless spores in the spring of the year. Infected pines as well as currant and gooseberry bushes within a radius of about 1,000 feet should be removed. Other measures that have proved effective are to refrain from making pure plantation of white pine and planting it in mixture with other species. The author has consistently advocated the planting of Norway pine instead of white pine in the Hudson Valley for over twenty-five years. Where white pine is native constant vigilance must be maintained to locate foci of infection.

Fungi Attacking the Wood

Wood-rotting fungi, which totally destroy the timber value of trees, certainly rank with the most devastating diseases of forest trees. Usually they attack only the heartwood of older trees, especially the extremely old or overmature ones, hence in forest management they could be kept in check by cutting trees before they reach this stage. Broken branches and injuries caused by fire, insects, frost and lightning supply the principal infection points. The elimination of diseased trees by felling and burning is too expensive except perhaps in timber-sale areas.

Among this large group of fungi three should be mentioned as fairly well representing the destruction caused by the whole group. Red rot or *Trametes pini* attacks many of our conifers. In the Northeast it is found on the spruces, white pine, hemlock, the balsam fir, and the larch. From the point where the spores enter, the fungus develops in the wood and spreads up and down the trunk, without showing any outside manifestation of its presence. After a number of years, fruiting bodies with a reddish-brown color on the under surface appear in the crevices of the bark, at old knot holes, and on the stubs of dead limbs. By this time the heartwood of the tree has been riddled with tiny holes which have a characteristic shiny, white lining. Sooner or later the tree is blown down or broken off near the base. *Polyporus Schweinitzii* is a destructive root rot of conifers which attacks living trees of all ages. The fungus attacks the tree through the roots underground and spreads to the lower part of the trunk. It spreads from one tree to another, and hence diseased trees are found in groups. Some in the group may have been uprooted or broken off. The lumber value of the heartwood is destroyed since the wood turns yellow and powdery when dry. Outward symptoms of attack are the reddish-brown fruiting bodies or brackets usually found in groups near the base of the tree. White rot or *Fomes igniarius* is one of the many heart-rot fungi found on deciduous forest trees such as poplar, aspen, maple, birch, and beech. It develops much the same way as the other rots and has conspicuous fruiting bodies which grow year after year to great size. They are reddish-brown, and many can be found on one tree. The wood is rendered soft and friable and hence trees are easily broken off by windstorms.

One of the most recently introduced fungi is the Dutch elm disease which threatens to destroy our elms as the chestnut blight has our chestnut trees. It was first found in Ohio in 1930 and has since been found as far west as Indiana and as far east as Connecticut. The elm bark beetle seems to be the cause of its rapid spread. The fungus kills the tree because it works in the sapwood and shuts off the water supply to the leaves which thereupon wilt and dry up. The Dutch Elm Disease Laboratory has been established by the federal government in New Jersey to diagnose the disease. Trees in the neighboring states known to be infected are removed and burned.

Destructive Soil Fungi

Undoubtedly one of the most pernicious soil diseases is that known as damping-off. It is a universal trouble affecting tree seedlings from the time of germination until the stem tissues become woody—a period of from one to two months depending upon the species. These fungi have no restricted host range and are usually associated with decaying plant material. The seedlings of coniferous trees show a more marked susceptibility than do those of deciduous trees, although beech and maple seedlings sometimes suffer greatly. Seedlings affected wither and die.

These fungi exist ordinarily upon decaying vegetable matter in the soil and their parasitism is very primitive. Their mycelium (delicate vegetative threads) is the main infective material. The spores rarely infect healthy plants, but they serve mainly to carry the fungus over the winter. The fungi live for long periods, probably indefinitely, if conditions of temperature, moisture, and food supply are favorable. Once a soil is infected it is likely to remain so for a long time and this infection is easily carried from one place to another. Control methods have been worked out and consist mainly of the disinfection of the soil by certain chemicals.

The Mistletoes

Several species of parasitic flowering plants, belonging to the family *Loranthaceae,* known as the mistletoes, cause considerable damage to forests and are probably the most widely distributed parasitic enemies. Very few genera of trees escape their destructive effects. Two kinds of mistletoes grow on forest trees: one belonging to the genus *Phorodendron,* the other to the genus *Razoumofskya.* The former are the large mistletoes common in the East, West, and South; the latter are the dwarf mistletoes which do such enormous damage to conifers in the West. Both groups materially reduce the growth of stands and kill trees outright. Over eighty species are known.*

Spruce, pine, larch, fir, and hemlock are attacked by more or less restricted species of dwarf mistletoes. These plants are very small and inconspicuous, since they lack leaves. When the mistletoe seed germinates, root-like sinkers are pushed down into the wood of the branch. The young mistletoe plant then obtains its water and food by robbing its host. The damage done by the dwarf mistletoes is largely due to the stimulus reactions set up. Excessive growth occurs in the part of the tree attacked and large numbers of short, bushy branches are produced which form what is commonly known as witches-broom. These brooms are often very large, and the excessive growth results in a lessened vigor of the tree and diminished annual growth.

The common holiday mistletoe has larger plants, and since it has green leaves is able to make part of its food. Like the dwarf mistle-

* W. H. Rankin, *Manual of Tree Diseases* (New York, 1918), pp. 54-60.

toes they impair the vigor of the tree and cause diminished growth. Sometimes the growth of this parasite is so dense that it cuts off the light supply to the leaves and the host tree dies. When a forester has a chance to combat these parasites in lumbering operations he can largely eliminate them by cutting out the infested trees.

Fungi Attacking Leaves and Twigs

Another group of fungi to be considered are those attacking the foliage and twigs of forest trees. While wood-destroying fungi affect our present stock of timber, the needle, leaf, and twig diseases affect our future crop because they interfere with the production of sound merchantable timber. These diseases are more dangerous to immature trees than to mature ones. If only a few leaves or branches are killed, others replace them or their functions are taken over by those that remain. When, however, a large part of the crown of the tree is damaged, the remaining leaves are not able to assume the whole work and the equilibrium between foliage and root system is seriously disturbed. Growth is retarded and the tree may eventually die. If this should happen before the tree is mature, the loss in expected timber would be considerable. Should an old tree die from the attack of these fungi, on the other hand, the damage done from a commercial point of view consists only in the cessation of growth. The timber already formed is available unless the tree is left in the woods to be destroyed by secondary fungi.

Diseases of leaves and needles are due very often to the interference of external factors with the physiological functions of the tree. Extremes of moisture and temperature, atmospheric impurities such as smoke and smelter fumes, and an impaired or improper supply of raw food materials, all lessen the vitality of the tree and lead to many leaf and needle diseases. Coniferous leaves may develop disease through winter drying. Late frosts in the spring may similarly affect both conifers and deciduous trees. Periods of drought, sunscorch, and excessive moisture may develop diseases. In most cases the leaves wither and die and later turn yellow and then red-brown in color. A new crop of leaves often replaces the ones so lost and the only loss is in decreased annual growth.

Besides these so-called physiological diseases, there are fungi which do considerable damage to foliage and twigs. These consist of leaf-spots, blights, mildews, leaf-casts, sooty molds, parch blights, and many others. Leaf-spot diseases are characterized by the formation of dead areas in the tissue of the leaf. They affect all kinds of deciduous trees. The leaves and sometimes the twigs and fruits of many deciduous trees are attacked by fungi which cause powdery mildews. Several diseases causing similar effects on pine, larch, fir, spruce, and juniper are known under the general name of leaf-cast. This causes the leaves or needles to turn reddish-brown and die. Molds cover the leaves of many trees with a black, sooty coating in mid-summer and this often becomes serious enough to interfere with the functions of the leaf. A disease called silver blight occurs on ash, chestnut, and other trees. The leaves and twigs at first become pale in color and then turn milky white.

Leaf diseases are important in forest production principally because they affect the organs that manufacture the food of the tree, and with this function impaired, the tree is unable to make as much annual growth as it would under normal conditions. If these diseases spread over large areas the total loss of wood growth may be considerable. It is not very often that leaf and needle diseases become serious enough to cause the death of trees.

Fungi Attacking Cut Wood Products

The last group of fungi to demand attention are those that affect forest products after they have been cut. Deterioration of these products is due to stain and mold organisms which discolor the wood but do not effect its structure or strength, and to the decay fungi which attack the cell wall material of wood and weaken, soften, or completely destroy the wood. Both groups of fungi grow only in moist wood but not in water-logged wood (hence the practice of storing logs in ponds). Such sap stains cause great loss in decreasing the marketability of pine, gum, and other species. Sap stain is especially troublesome in sawlogs delayed in getting to the sawmill and in lumber during seasoning. Prompt milling would cure the former trouble. Lately logs have been given a surface spray to prevent fungous growth. As for lumber, kiln-drying is a preventative for the higher grades of lumber, and an inexpensive antiseptic spray has been developed for the lower grades. Losses from decay are due to forest products being stored in the woods or at sawmills, or to conditions of actual use that favor decay. These losses can be reduced greatly by shortening the storage period and by employing modern wood-preservation methods on ties, poles, posts, piles, house timbers, and other products used in or near damp soil.

Pathological investigation has brought out a number of arresting facts about tree diseases, which threaten to become a very important limiting factor in growing seedlings in nurseries, in establishing plantations, and in reproducing forests naturally. In nursery practice root rots, molds under snow cover, and stem rusts may cause the abandonment of certain species in certain localities. The brown-spot needle disease of long-leaf pine in the South is preventing the establishment of young stands of that species in certain areas. The blight of spruce and fir in the Northeast presents a similar problem in that section. Even exotics like Scotch pine are being attacked by rusts. It is very evident that planting sites must be chosen with reference to disease hazards and that soil types must be evaluated on the same basis. Then again the avoidance of pure stands may help to lower the hazard. Some trouble has been experienced with species planted outside of their natural range. Finally, it becomes plain that forest-tree geneticists will have to develop immune varieties for some forest trees just as the plant geneticist has bred immune types of wheat, corn, potatoes, and many other crop plants.

Investigations of forest pathologists have also given us estimates as to the amount of loss due to fungi. Our most reliable data deal with over-mature stands of Douglas fir where the loss from decay fungi in merchantable stands is about 17 per cent. This figure is accurate, because it is based on actual cull experienced in lumbering operations. Estimated cull in standing saw timber is given as oak 19 per cent, birch, beech, and maple 21 per cent, eastern hemlock 18 per cent, western hemlock and western red cedar 20 per cent. In certain sections of the country cull runs even higher. Western hemlock in northern Idaho, the white fir and incense cedar of the far West, beech in the Adirondacks and in the bottomlands of Louisiana are often largely unmerchantable on account of heart rot. It has been found that decay increases with age and hence could be controlled by short rotations. Above certain ages, the increase in decayed wood exceeds the total increase in tree volume for the same period. Decay is said to exceed growth in the over-mature stands of Douglas fir, white fir, western white pine, incense cedar and others. Furthermore, the losses due to dwarf mistletoes and similar diseases in immature stands probably equals the loss caused by decay in certain mature stands of western yellow pine.

A very important fact as to the cause of decay has been brought out. Decay fungi commonly enter through open wounds, and the most common wounds are fire scars. Decay in the lower part of the trunk (caused by frequent ground fires) destroys the most valuable part of the tree because it is these logs that yield the high grades of lumber. In eastern hardwoods most decay is directly connected with fire scars. Here is another extremely good reason for fire prevention.

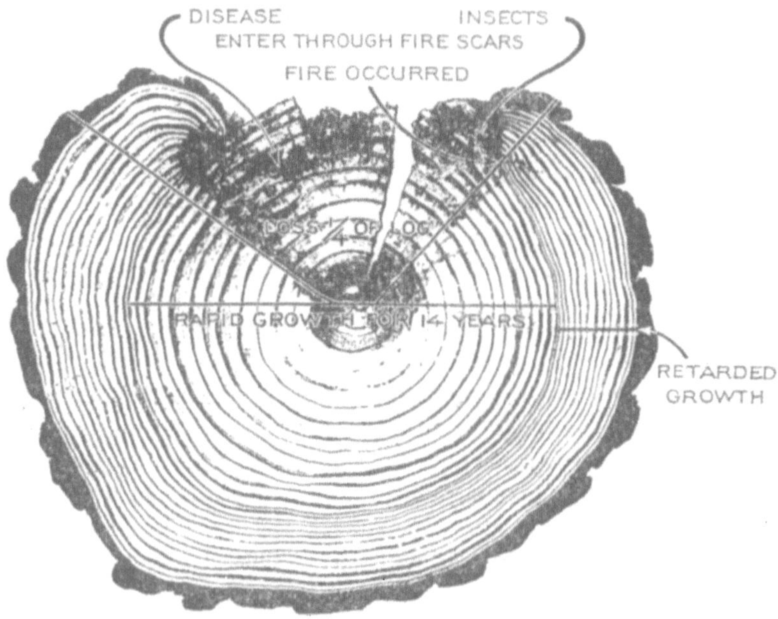

Cross-section of a tree showing fire scar (when 14 years old) followed by insect and fungus attacks, rendering one-fourth of the log waste. Notice the retarded growth following the fire.

Work in forest pathology is carried on by the federal government in the Bureau of Plant Industry, the Division of Forest Pathology, and the Division of Blister Rust Control. The Division of Forest Pathology is an investigative organization employing a number of forest pathologists to develop principles for control to be later applied by the Forest Service, the Division of Blister Rust Control, the National Park Service, state forestry departments, and to some extent by private owners. Besides this centralized service some pathological work is done in a few state forestry departments and at the state educational institutions and the agricultural experiment stations. New York, Illinois, Ohio, Pennsylvania, Michigan, and other states are doing individual work.

CHAPTER XIV

LEADERSHIP OF THE FEDERAL GOVERNMENT

HISTORY BROADENS OUR perspective, sharpens our judgment, and develops wisdom for inaugurating future policies. Leading men make events; events shape history. Nowhere is this better illustrated than in our national forestry movement. Dynamic, far-sighted, and unselfish American leaders have for more than sixty years blazed the trail towards the goal of sustained-yield management of our forest resources. It makes an interesting story.

We can understand the development of federal forestry better if we keep in mind some of the generally accepted ultimate objectives of the federal government in forest conservation. Even though the final goal is still a long way off it is interesting to see how we are approaching that goal step by step.

The federal government will, no doubt, always be the largest single owner of forest property in the country. Its holdings will therefore be important primarily for watershed protection, and secondarily, for safeguarding the timber supply of the nation. Federal holdings will always give that much-needed stability to the development and the utilization of the vast store of raw materials found in the forest. They will give insured permanency to such great American industries as lumbering, wood-using establishments, grazing, agriculture, recreation, and water power, not to mention such vital services as public health, erosion and flood control, domestic water, and many minor forest products. Its holdings should be largely confined to protection forests on the upper slopes of mountain ranges where soil and climatic conditions result in a slow rate of growth and a low rate of financial return on the investment. Such lands will never be attractive to the private investor. Such lands must, moreover, always be the concern of the federal government because of their protective functions which necessarily have interstate implications.

The federal government would naturally, from its strong financial

Photo courtesy E. I. du Pont de Nemours and Company

RAYON FILAMENT FROM LIQUID VISCOSE

MOST OF OUR PAPER IS MADE FROM WOOD

This modern machine turns out ten-foot-wide ribbons of paper at the rate of a mile every five minutes.

U. S. Forest Service Photo

U. S. Forest Service Pho:

YEARLING STEERS ON CUT-OVER MEADOW, MONTANA

CATTLE LEAVING BEAVERHEAD NATIONAL FOREST FOR THE HOME RANCH, MONTANA

U. S. Forest Service Pho:

U. S. Forest Service Photo

SHEEP GRAZING IN GALLATIN NATIONAL FOREST, MONTANA

COUNTING SHEEP AS THEY ARE RETURNED TO SUMMER RANGE

U. S. Forest Service Photo

Photo U. S. Soil Conservation Service

SEVERE EROSION IN APRICOT GROVE, VENTURA COUNTY, CALIFORNIA
This represents a soil loss of over 400 tons an acre.

ROAD AND FARM THREATENED BY EROSION
View on Salt Fork River, Oklahoma. The trouble starts with abuse of land and forests at the head of the river.

Photo U. S. Soil Conservation Service

Photo U. S. Soil Conservation Service

STRIP-CROPPING OF WHEAT AND COTTON, NEAR SPARTANBURG, SOUTH CAROLINA

This practice holds the soil from erosion.

CONTOUR PLOWING AND STRIP-CROPPING, WEST VIRGINIA

Photo U. S. Soil Conservation Service

U. S. Forest Service Pho

GREAT FIRE IN FLATHEAD NATIONAL FOREST, MONTANA, 1929

U. S. Forest Service Photo

FIRE EATING AGAINST A STRONG WIND

Spud Hill fire, 1937, Columbia National Forest, Washington. Forest fires upset our national economy not only through loss of natural resources but also through the extra burden on taxes, and the loss of employment and payrolls. It takes 80 to 100 years to replace a commercially valuable forest.

FIRE-SCARRED WASTE, SIUSLAW NATONAL FOREST, OREGON

U. S. Forest Service Photo

U. S. Forest Service Ph

CONSTRUCTING FIRE LINE, GEORGE WASHINGTON NATIONAL FOREST, VIRGINIA

U. S. Forest Service Photo

FIGHTING FOREST FIRES IN MASSACHUSETTS

In 1937 there were over 185,000 forest fires in the United States, burning over 22,000,000 acres. Massachusetts is one of the better-protected states.

U. S. Forest Service Photo

FIRE TRUCK DISTRIBUTING EQUIPMENT
This crew is stopping a fire in the Sumter National Forest, South Carolina.

SILT-FILLED RESERVOIR: A RESULT OF FOREST FIRES
This shows what happened in Spartanburg County, South Carolina.

U. S. Soil Conservation Service Photo

U. S. Forest Service Photo

FIRE LOOKOUT, FRANCIS MARION NATIONAL FOREST, SOUTH CAROLINA

U. S. Forest Service Photo

LOOKOUT TOWER, SNOQUALMIE NATIONAL FOREST, WASHINGTON

These towers are the "eyes" of the forest service. Note observer on catwalk, scanning the horizon for possible fires.

LOOKOUT TOWER, HOLLY SPRINGS NATIONAL FOREST, MISSISSIPPI

These pine lands are so level that high towers must be erected to afford a view over a wide area. Note the substantial administration buildings.

U. S. Forest Service Photo

U. S. Forest Service Photo

FOILING GYPSY MOTH CATERPILLARS

Since the introduction of the gypsy moth from Europe in 1869, damage done to trees by this insect has amounted to many millions of dollars.

U. S. Forest Service Photo

ADMINISTRATION BUILDINGS, DEEP CREEK RANGER STATION, BITTERROOT NATIONAL FOREST, IDAHO

LOGGING WITH POWER EQUIPMENT, IDAHO NATIONAL FOREST

Photo by Sibbold, courtesy U. S. Forest Service

U. S. Forest Service Photo

MEASURING LOGS, PAYETTE NATIONAL FOREST, IDAHO

U. S. Forest Service Photo

WPA WORKER PLANTING SHORTLEAF PINE

Member of crew working in Hoosier Purchase Unit, Indiana, 1940.

MARKING MATURE TREES FOR CUTTING, COCONINO NATIONAL FOREST, ARIZONA

U. S. Forest Service Photo

position, be the agent most able to undertake alone or coöperatively large-scale reforestation projects, whether these be cut-over lands in the Lake states or the South, burned-over barrens in the northern Rocky Mountains, or the chaparral fields of California.

The federal government should always take an active lead in co-operation with states and groups of states, and through them with private individuals in such matters as fire protection, reforestation, management, acquisition of state forests, taxation, insurance, water-shed protection, and other problems of an interstate character. Through its coöperative laws it can insist on uniformity of method and on minimum requirements to assure efficiency.

Lastly, the federal government should be the balance wheel in the forestry movement and act in an advisory capacity to states and to private owners for the good of all the people and not merely for the benefit of certain states or groups of states. It should never monopolize the situation. There will always be a definite place for state forestry, community forestry, and private forestry. The federal government should never undertake to do what these smaller owners can do as well for themselves.

The Rôle of Government in Forestry

We believe that the entire future forestry policy of the United States must be based on the assumption that forestry, *per se,* is largely a governmental function. Eventually, over 50 per cent of our forest lands probably will be administered by state or federal government. The nature of the forestry business makes this inevitable.

Since profit is the basic incentive of private enterprise, exploitation of this natural resource is too apt to be wasteful with far-reaching and dangerous consequences. Having removed all marketable material, in most cases the private owner tends to lose interest, and the area is subject to the inevitable consequences of fire, which soon render it a barren waste, incapable of producing any economic return. Not only are the young growth and soil destroyed but the consequent and cumulative damage often extends far beyond the owner's property to the farms in the valleys below and to the cities and towns in other states. Of course, the private individual cannot be expected to endure the economic burden of managing his property at a loss to himself, even if by so doing he is protecting the interests of distant communities. Hence the government must step in either by making financial conditions favorable for continued productivity, or by taking over the property and assuming the entire responsibility in accordance with the welfare function of government.

Furthermore, management for continuity means the sacrifice of present revenues and the making of present expenditures for the sake of greater returns fifty to a hundred years hence. Only a long-lived, wealthy, and stable class of ownership can afford to make such sacrifices. The leaving of an enormous wood capital and large areas of immature growth on the land after logging can be carried out only by authorities which can cope with the all-important time element in forest management. The federal, state, and local governments, by their very nature, are best able to do this.

The history of forestry shows that the rôle of government is at first merely persuasive and educative. Economic and industrial development in a country may later require restrictive measures and the exercising of police functions. The final stage which has come about in many European countries, evidently because the other two methods have failed, is that of direct government ownership and management.

History Repeats Itself

A technical art like forestry, which relies so much on experience and practice, must give attention to historical precedent. It is necessary to review in proper perspective the accumulated experiences of the past so that we may better understand our present situation and plan our future course with some degree of assurance. In modern parlance, we should find out how we got this way and then determine what we are going to do about it.

The most obvious lesson to be learned from a study of the history of forestry is that history repeats itself. At first the forest had little value. The supply seemed limitless; the demand of the meagre population was inconceivably small compared to the supply. The forest was even regarded as an undesirable incumbrance of the soil. It was often destroyed to make way for pastures and farms. Land had to be cleared, for the people had to be fed.

Increasing population accelerates forest destruction. This generally ushers in an era of apprehension. One hears the cry to protect the forests from lumbermen, forest fires, and cattle. Restrictive measures are first inaugurated against forest fires. The first really constructive efforts appear in the form of tree planting laws and the protection of natural reproduction. It is the beginning of silviculture, for the people are wondering where the next crop of trees is coming from. Continued economic and industrial expansion brings the era of threatened exhaustion of virgin supplies. Then follows the government acquisition of forest lands and their management for continuity, with eventually a certain amount of government regulation of private forest lands.

Our Public Domain

Since a large part of our federal government's forest policy is intimately related to the disposition of its public domain, it is necessary to consider briefly how this was acquired and what was done with it.

By 1853 the federal government, through treaties and purchases, found itself in possession of about one-and-a-half billion acres of land. This comprised roughly the vast expanse of territory west of the Mississippi River, a region of great potential wealth. In fact it is doubtful whether any similar sized area in the entire world could display such a wealth of forests, water power, oil, coal, and mineral deposits. Settlement had scarcely begun. Hence the government embarked upon one of the most lavish systems of land disposal ever attempted by any nation. This system was based upon the belief that only through private ownership could the settlement and development of the West be encouraged. Furthermore, it was expected that the sale of these lands would finance part of the cost of government. A few statesmen like John Quincy Adams had foreseen the danger of this policy—but the doctrines of Andrew Jackson had become predominant.

During the next forty years over a billion acres of valuable land passed to private ownership. Many land laws were enacted to encourage agricultural settlement. Military scrip laws, the homestead law, the timber culture law, the timber and stone law, and the desert land law were passed to encourage private ownership. In addition vast grants were made to individuals and corporations to aid in the construction of wagon roads, waterways, and railroads. The railroads alone received over 130,000,000 acres of free land. The public land states were given immense areas of the domain for public education. Over 60,000,000 acres of swamp lands were granted to states in expectation that the land would be reclaimed by drainage.

The question is here pertinent: Did the government's system of encouraging private ownership work? It soon became apparent that the original purposes of the laws and acts of Congress were being fraudulently evaded. This applies to states as well as individuals. People secured claims under these acts merely for the timber or other wealth that was on them. The government had no administrative and protective organization to enforce either the letter or the spirit of the laws. The resulting squandering of our natural forest wealth is one of the darkest pages in our national history. It is true that we secured the settlement and opening of the great West, but it was all done at a tremendous sacrifice.

By way of contrast, consider our neighbor, Canada, which developed its western provinces but did not give away its great forest wealth in so doing. Even today over 90 per cent of the forest land of Canada is owned either by the Dominion or by the provincial governments, and the timber is sold under leases.

It might be said, parenthetically, that the mere remnants of our vast original public domain are today found in our National Forests, National Parks, National Monuments, Indian lands, reclamation withdrawals, and lands set aside for their coal and potash value. The area thus set aside for the people's welfare aggregates about 250,000,000 acres. Then there are approximately 160,000,000 acres of unreserved and unappropriated federal land, most of which is good only for grazing, and of which about 2.5 per cent might be classified as commercial timberland.

Early Colonial Legislation

While this dissipation of the nation's timber wealth was going on, the national conscience gradually became awakened to the need of a more rational use of our forest resources. This movement finally culminated in the Forest Reserves Act of 1891, which attempted to save what was left.

When the colonists landed at Jamestown, Cape Cod, and other points along the Atlantic Coast, unlimited forests confronted them. Clearing of the woods to make room for farming began immediately. The first commercial sawmill was built in the town of Berwick, Maine, in 1631. Not many years later, ships loaded with masts, spars, clapboards, shingles, and large ship timbers were clearing the ports of New England for Europe. For many years the timbers for the finest ships of the British Navy came from New England.

Slowly the forests were pushed back. With the opening up of western Pennsylvania, Ohio, and adjacent territory, more agricultural lands were cleared. But the real impetus to lumbering did not arrive until the settling of the Middle West. By 1840 there were over 30,000 sawmills, one-third of them in New York and Pennsylvania. Up to 1850 Maine led all states in lumber production. In that year New York gained the lead, followed in 1860 by Pennsylvania. Thirty years later the Lake states reached the peak of their production.

Early fragmentary regulations and ordinances were dictated not by any threatening timber shortage, but merely to secure the proper use of public property. As early as 1626 the colonies and the town governments of New England passed laws to protect their forests. William Penn, the founder and first legislator of the state which bears his name, in 1681 stipulated to those who took title to land from his grant, that for every five acres cleared, one should be reserved for forest growth. It was a wise provision but it unfortunately bore little fruit.

The early colonial ordinances remind one of the restrictive laws of western Europe in the fifteenth and sixteenth centuries. As in Europe, the colonies early focused their attention on fire protection. As early as 1743 a Massachusetts act recognized the damage caused by fire to forest reproduction and to the soil, and in 1777 a North Carolina law penalized the unlawful firing of the woods and recognized that forest fires are extremely harmful to the soil.

The first legislation in America which was dictated by the threat of timber exhaustion, was an act approved in 1799 providing for the purchase of a forest reserve for ship timbers for the Navy. Under a series of acts a total of 244,000 acres of forest land was reserved in Florida, Mississippi, Alabama, and Louisiana for the growth of live oak, cedar, and other timbers suitable for the use of the Navy. Laws were also passed for the protection of these reserves. It is certainly to be greatly regretted that this legislation was not followed up by a broader policy of forest reservation in other states and for other purposes. About sixty years elapsed before this policy was revived.

Post-Civil War Period

It was not until after the Civil War that the next progressive steps were taken. As was to be expected, the states in which forest destruction had reached its worst stages were the first to attempt to mend their ways, thus leading the way along which the federal government was soon to follow.

As we have already seen, the decade following the Civil War was marked by the construction of our great trans-continental railroads and the consequent development of the western states. Between 1865 and 1875 the railroad mileage of the United States doubled. The government land policy just described greatly encouraged construction and subsequent settlement. Towns and villages sprang up like mushrooms. Since the prairie states had no timber of their own, it was natural that the destruction of the forests of the Lake states received added impetus.

The first attempt to remedy the situation was made by the states of Michigan and Wisconsin in 1867 when their legislatures passed acts providing committees which should report on the destruction of the forests of those states to see what could be done about it. In a similar manner Maine considered a forest policy for the state when the State Board of Agriculture appointed a committee for that purpose. From then on until 1891 seven other states (New York, Connecticut, New Hampshire, Vermont, Ohio, Pennsylvania, and North Carolina) appointed similar commissions of inquiry to look into the condition of their forest resources. However, very little of permanent value resulted.

At about the same time a wave of persuasive legislation developed, some of which had some influence in developing a better appreciation of our trees and forests. Between 1861 and 1871 Iowa, Kansas, Dakota, Nebraska, Minnesota, Missouri, and Wisconsin passed laws to encourage tree planting. In most cases the inducement was tax exemption or bounties. In 1872 Nebraska inaugurated Arbor Day for planting trees and this idea spread to practically every state in the Union. Arbor Day is now a national institution and it certainly has had a far-reaching educational effect, although its immediate value to the cause of forestry has been negligible.

So far there were no federal measures of any value to be recorded. There was rapidly developing, however, an undercurrent of public opinion so indispensable in any great national movement. Sooner or later this must bear fruit. It was difficult to make the people believe that our forest wealth was not inexhaustible, or to feel the necessity of urging legislators to take action.

A few men of vision, however, saw the importance of saving our timber resources and unselfishly gave of their time and energy to get the forest conservation movement started. One of the first of these pioneers was André Michaux, a French naturalist and traveller. He made some very profound and incisive remarks in his *The North American Sylva,* which was published in 1819. Being perfectly familiar with the lessons learned from forest destruction in his native France, he was alarmed at the thought that neither the states nor the national government had taken any action. Among other things, he left a legacy to endow a series of forestry lectures. These were given at the University of Pennsylvania by Dr. J. T. Rothrock, Professor of Botany there from 1877 to 1893. Dr. Rothrock later became Commissioner of Forestry for his native state, 1893 to 1905, and is generally called the "father of forestry" in that state.

Another book worth noting at this time was G. P. Marsh's *Man and Nature* (later revised under the title *The Earth as Modified by Human Action*), which considered at great length the effect of forest destruction on climate, rainfall, erosion, and floods. These facts were no doubt impressed upon the author while he was ambassador to Italy. The book was first published in 1868.

Dr. F. B. Hough is often called the "father of American forestry." He was a physician, but like so many of his profession he delved into many fields. He was also a writer, historian, botanist, mineralogist, statistician, and student of forestry. In 1873 he presented a paper at the Portland, Maine, meeting of the American Association for the Advancement of Science entitled "The Duty of Governments in the Preservation of Forests." This led to a forestry committee of the association memorializing Congress to create the office of Commissioner of Forestry to gather forestry information. As we shall see, this bore fruit in 1876.

The first popular forestry association was organized in St. Paul in 1875 and this became in 1882 the American Forestry Association. This organization has always been the center of private forestry effort in the United States. Much of the early propaganda work was done through this organization and its journal, now called *American Forests.*

First Federal Legislation

The first important step taken by the federal authorities was at the urgent request of the statistician of the Department of Agriculture in 1870. At that time lands were recognized as being either "improved" or "unimproved" farm lands. He recommended that in the next census the category of "unimproved farm lands" be further subdivided into "woodlands" and "other unimproved lands." By this classification more attention was concentrated upon the woodlands. Investigations shortly thereafter showed that 39 per cent of the area of the country was in woodland. This was the first and most logical step toward taking an inventory of our forest resources.

Another early attempt by the federal government to assist in forest conservation was the encouragement of tree planting on the plains of our western states. The legislatures of several of these states had already passed a number of laws in this direction. On March 3, 1873, the Timber Culture Act was passed by Congress, by which the planting to timber of 40 acres of land in the treeless prairies conferred the title to 160 acres of public domain. The act proved a dismal failure because settlers had no knowledge of tree planting, the restrictions of the act could not be enforced, good planting material was not available, and the act was open to other abuses. It was finally repealed in 1891.

The first constructive piece of legislation enacted by the Congress of the United States was the Act of August 16, 1876, which provided for the appointment of a Commissioner of Forestry (actually called an "Agent") who should ascertain facts applicable for the preservation and restoration of forests and transmit the same to the Commissioner of Agriculture who should transmit the report to Congress. As already mentioned, Dr. Franklin B. Hough was the first man to be appointed under this act. He prepared several voluminous reports on forestry conditions which were laid before Congress. But the appropriations during these years were so meagre that little could be done. The first forestry appropriation was for $2,000, a proviso attached as a rider to the free-seed clause in the Legislative, Executive, and Judicial bill for 1877. In 1878, $2,500 was appropriated; the next two years no money was appropriated; and in 1881 the amount grew to $5,000. N. H. Eggleston succeeded Dr. Hough in 1883 and served until 1886.

Another personality that stands out in these drab years is that of Carl Schurz, Secretary of the Interior. His department had charge of the forests on the public domain, and from 1877 to 1881 he staged a gallant fight for the correction of abuses and the establishment of a far-sighted forest policy. Lack of popular support and the idea that public forests were to be exploited for private gain frustrated his efforts. He fought and lost, but he will always be remembered as the first champion of our present forest policy.

In 1886 a distinct bureau, the Division of Forestry, was established in the Department of Agriculture. This was actually the first statutory provision for forestry by the federal government. Dr. B. E. Fernow, a technically trained forester, was made Chief. The only thing we can say for this change, however, is that it added to the dignity of forestry in the federal government. The duties and functions of the Division were slow to expand. Prior to this date, in 1881, an agent of the Department was sent to Europe to study the practice of forestry there. In 1884 the duty of making experiments with timber was added to the functions of the Division, and the next year the collecting and distributing of economically important tree seeds were begun. In 1886 the study of the biology of some of our important timber trees was taken up, while the following year silvicultural problems first engaged the attention of the Division.

Forest Reserves Act

While all this preliminary work was being done, not an acre of the vast forested public domain had been set aside for the use of all the people. Various land laws and grants were taking the cream of the valuable timberland out of government ownership. Such restrictive and protective laws as there were could not be enforced on account of the lack of appropriations to maintain an administrative and protective organization. The time was now ripe for an executive policy to manage the woodlands that still remained in the possession of the government before it was too late to save what remnants were left.

Before 1891 the Division of Forestry was simply a bureau of information. In general the information supplied was of a twofold nature. It was technical in so far as it related to the management of private timberlands and statistical in so far as the knowledge of the conditions of our forest resources induced the application of forestry principles. Up to that date Congress had neither appropriated enough money for efficient outdoor work nor attempted to put any government woodlands under the control of the Division. There had been no forest management because there were no statutes which would permit such management.

But the Division of Forestry was designed by the nature of its duties to be more than a bureau of information. The existence of a governmental department to promulgate forestry principles while the government itself had made no provision to apply such principles to its own permanent timberlands was an incongruity that demanded further legislation. This was in part supplied by the Forest Reserves Act of March 3, 1891, which conferred upon the president the power to establish Forest Reserves from the public domain. It is interesting to note that this act was not a deliberate piece of legislation openly discussed in both Houses of Congress, but it was slipped in as a "rider" at the last hour in Conference Committee.

The first exercise of power under this act was the proclamation by President Harrison creating the Yellowstone National Timber Land Reserve on March 30, 1891. This was probably the most constructive step yet taken in the building of a National Forest policy; but, unfortunately, the act left the Division simply a bureau of information as it was before, because it made no provision for the administration or protection of the reserves, nor for the practice of forestry.

The very reasons for establishing the Reserves, as stated in the law, would point to the need of some kind of a system of administration to make the law effective. Some of the reasons given were: to prevent annual conflagrations, to provide revenue from the sale of forest products and timbers, to increase the stock of game, to promote the development of the country, and to get the maximum good from each acre. Furthermore, the law withdrew from public use all such lands that might be acquired under it. It really meant that the forest resources were locked up so they could not be used.

In spite of the anomalous condition of forest reserves without forest administration, more forest reserves were created by Presidents Harrison and Cleveland. By the close of 1897 a total of almost 40,000,000 acres had been reserved. They were placed under the jurisdiction of the General Land Office. As appropriations for the Division of Forestry (Dept. of Agriculture) were small, amounting to less than $30,000 annually, little could be done. This money was used mainly for testing timber strength and the conditions affecting quality.

First Administrative Attempts

The Secretary of the Interior in 1896 requested the National Academy of Sciences, the legally constituted adviser of the government in scientific matters, to investigate, report upon, and recommend a national forest policy. This resulted in the Act of June 4, 1897, under which, with subsequent amendments, the Forest Reserves are now being administered. Under this act the Reserves remained in the hands of the General Land Office, Department of the Interior. It charged this office with the administration and protection of the Reserves. Later the Geological Survey was charged with surveying and mapping them, and the Division of Forestry of the Department of Agriculture was asked to give technical advice. It is very evident that the latter, containing the trained scientific staff, had no relation to the government forestry work, except as the offices of the Department of the Interior might apply for assistance and advice. This complete separation of the administration of the General Land Office and the force of trained men in the Division of Forestry was a serious defect.

However, in spite of this tripartite control some progress was made. The U.S. Geological Survey undertook the task of surveying, classifying, mapping and describing the Forest Reserves. At a cost of about one-and-one-half million dollars over 70,000,000 acres of Forest Reserves were mapped and described. The General Land Office undertook the administration, and forest superintendents and rangers were appointed to take charge of the Reserves. The rules and regulations for administration were formulated by the Commissioner of the General Land Office. Filibert Roth compiled the first "use" book.

A good idea of the limited extent of the forestry work of the federal government at this time can be secured by a brief glance at the personnel and activities of the Division of Forestry in 1898. On July 1 of that year it employed eleven persons, six clerical and five scientific. There also were some collaborators and student assistants. There was no field equipment and little field work. But in the fall of 1898 an important step was taken. From that time on the Division offered practical assistance to forest owners, and this shifted its field of activity from the desk to the woods. One of the first working plans made was for a private estate in the Adirondacks. The lumbermen were met on their own grounds, and actual forest management for purely commercial ends was undertaken by well-known timberland owners. From that time dates the solution of specific problems of forest management. The work of the Division at this time, therefore, consisted of activities along four distinct lines: (1) working plans, (2) economic tree planting, (3) special investigations, and (4) office work. Still it must be noted that the Division had practically nothing to say about the scientific forestry methods which should be used on the Reserves.

In 1901 the Division of Forestry was raised to the rank of a Bureau, but this was a change in name only and carried with it no change in handling the vast Forest Reserves.

Great Names in Forestry

We now reach the year 1905, almost thirty years after the first appropriation for federal forestry. Yet with all the varied enactments and laws, very little of permanent value had been accomplished. It might be interesting to search for the reasons behind this delay in getting forestry established and to point out briefly the pioneers who broke the deadlock.

Here are some of the reasons why a federal forestry policy was so slow in getting established:

1. Forestry was a new idea and it was hard to overcome established custom and prejudice, especially in changing over from a method of wholesale squandering to one of restricted and regulated use.

2. In not a few cases the movement was made a political football by self-seeking politicians. "Forest Conservation" was a new catchword for the gullible public.

3. Arbor day and tree planting sentimentalism often discredited and beclouded the real issue. For this reason lumbermen and others misconstrued the purpose of the reform movement.

4. The principal handicap, no doubt, was lack of educational facilities and trained foresters to instruct legislators and the general

public. The movement lacked information based on forestry facts; sentimental propaganda made only a temporary emotional appeal.

Educational facilities were slow in developing. Some talks and lectures on forestry were given as early as 1873 at Yale and 1874 at Cornell. Usually such lectures were given in the college Department of Botany. One of the earliest of these non-professional courses in forestry was given at the University of Michigan from 1881 to 1884 by Prof. V. M. Spalding of the Botany Department. The Michaux lectures at the University of Pennsylvania by Dr. Rothrock have already been mentioned. By 1897 over twenty land grant colleges offered courses in the subject, much of it being forest botany.

The first technical collegiate instruction in forestry was offered by Cornell in 1898, at Yale in 1900, and at Michigan in 1903. Dr. Schenk's private school at Biltmore, N. C., was opened in 1898 but closed in 1912. The Cornell school (N. Y. State College of Forestry) was discontinued in 1903, and reopened in 1910, and discontinued again in 1937. Five other important schools opened before 1905. By 1935 there were about twenty-five professional forestry schools in the country.

It will be recalled that the first two federal agents, Hough and Eggleston, lacked technical forestry training. The first technically trained forester was Dr. B. E. Fernow, who was Chief of the Division of Forestry from 1886 to 1898, after which he left to take charge of the first forestry school at Cornell.

Dr. Fernow received his technical forestry education in Germany and came to this country in 1876. For about forty years his many activities both in this country and in Canada made him known as the "Dean of American Foresters." He was not only an administrator and investigator, but also an educator, editor, and author. There is not a phase of forestry that has not been greatly enriched by his labors. He served as Dean of the Faculty of Forestry of the University of Toronto from 1907 to 1919.

While in the federal service Fernow had a close and sympathetic friend in Filibert Roth, who, besides doing pioneer work in timber physics, took a leading part in the organization of the administration of the Reserves in the General Land Office. "Daddy Roth," as he was affectionately called by his students, came to this country from Germany in 1871 and was largely self-educated, though he attended the University of Michigan from 1885 to 1893. He became a timber expert in the U.S. Department of Agriculture in 1893 and served until 1898, when he went to Cornell with Fernow. He entered the Bureau of Forestry in 1901 and became Chief of the Division of

Forest Reserves in the General Land Office in the same year. He was Professor of Forestry at the University of Michigan from 1903 until the time of his death in 1925.

Gifford Pinchot, who had studied forestry abroad, served successively as Chief of the Division of Forestry, Chief of the Bureau of Forestry, and Forester in the U.S. Department of Agriculture from 1898 until 1910. He had been forester for the Biltmore estate from 1892 to 1894 and had there instituted the first systematic forestry work in this country. When he entered government work he was young, ambitious, and influential, just the type of man that was needed at the time. In 1901, after the assassination of President McKinley, Theodore Roosevelt became President and threw his dynamic personality into the fight. Pinchot and Roosevelt used their enthusiasm and unlimited energy to advance the cause of forestry at the time when it was most needed. Pinchot later (1920-23) became Commissioner of Forestry for Pennsylvania and then Governor (1923-27, 1931-35). Over a long period Pinchot has been connected with the Yale School of Forestry, which was founded by a gift from his father, James W. Pinchot.

Adjustment of Administration

The foundation of federal forestry work may be said to have been laid by the Act of February 1, 1905. The necessity of consolidating the various branches of government forestry work became apparent and was urged upon Congress by President Theodore Roosevelt, Mr. Pinchot, and the executive officers concerned. By this act the entire jurisdiction of the Forest Reserves was transferred to the Secretary of Agriculture. Matters of surveying and passage of title, however, were still under the jurisdiction of the General Land Office, Department of the Interior. By this piece of legislation, the Division of Forestry for the first time in its career became an administrative organization. On July 1 of the same year the Bureau of Forestry became the Forest Service, and in 1907 the change in name from "Forest Reserves" to "National Forests" was made to correct the impression that the Forests were withdrawn from use.

Following the establishment of forestry schools at the beginning of the new century, the first professional foresters were taking upon themselves the task of applying forestry methods to the National Forests. Further evidence of the growth of the profession was the organization of the Society of American Foresters in 1900 and the establishment of the first professional journal in 1902, called the *Forestry Quarterly*. Since 1898 forestry knowledge gathered in the field has been forming a rapidly growing literature on the subject. Today good textbooks on the many phases of forestry science have been written, and what is most important, these are based on American instead of European conditions.

By 1905 the work of the Forest Service had increased to such an extent that the number of employees had increased to 821. Untrained, non-scientific woodsmen of the old type were being gradually replaced by trained foresters. In addition, the entire force was made a part of the classified Civil Service and the practice of making political appointments was apparently abolished. While the administration was being adjusted, the area of National Forests was also being constantly increased. To the 40,000,000 acres of Reserves set aside by Presidents Harrison and Cleveland, President McKinley added over 7,000,000 acres by 1901.

When Theodore Roosevelt became President, the National Forest policy received an added impetus and vigor. Being a great lover of the out-of-doors life and being especially well acquainted with the great West through residence (1883-87) and later through extensive travels, he threw his powerful influence into the balance for a cause in which he thoroughly believed. With the close coöperation of Mr. Gifford Pinchot, his warm personal friend and his Chief Forester, Mr. Roosevelt set aside between 1901 and 1909 over 148,000,000 acres of National Forests, or more than three times as much land as had been set aside by all his predecessors put together.

Since 1909 a careful adjustment of the boundaries has been going on, presidents since that time adding small areas here and there which were found valuable for forestry purposes, or eliminating small areas found to have no value. Acts of Congress passed since 1907 prohibit the addition by the president to the National Forests already established in Washington, Oregon, California, Idaho, Montana, Wyoming, and Colorado. Additions can be made in these states only by special act of Congress. A number of such acts have been passed already, some of them upon petitions of the peoples of the states involved.

Further slight modifications in the organization of the Forest Service were made during the Roosevelt and Taft administrations. Before 1908 all the work of the Forests was supervised from the main office in Washington, and this arrangement often caused much delay and inconvenience in carrying on the business of the Forests. In the fall of 1908 six administrative districts were established, to which another was added in 1914. Now there are ten. By this arrangement the National Forests are divided into ten groups and each has a regional headquarters in a large city or town centrally located in the group. The regional office acts as sort of a clearing house for all National Forest business. All matters concerning the administration and protection of the National Forests that cannot be decided on the individual Forest, or appear to be of general importance to the region, are taken to the regional office, which is in charge of a regional forester and several assistants.

Beginning in 1909 Forest Experiment Stations were established in each district, and in 1910 the Forest Products Laboratory, the first of its kind in the world, was formally opened in Madison, Wisconsin.

The Weeks Law of 1911

With the general administration of the National Forests of the West well established, the federal government saw a new and important duty loom on the eastern horizon.

For a number of years the need for government assistance in the southeastern states had become apparent. In the southern Appalachian Mountain region, floods and erosion as a result of forest devastation were making their appearance to an alarming extent. America's great hardwood forests were being decimated, and the problem of government control was difficult because the timberlands were entirely privately owned and there was no law which would permit the acquisition by purchase of private land for National Forests. The passage of a purchase law would certainly greatly strengthen the principle of direct government ownership and management.

Such a law came with the passage of the Weeks Law on March 1, 1911, which provided for the acquisition of National Forests in the East and the creation of the National Forest Reservation Commission, empowered to purchase such areas at the headwaters of navigable streams. The measure likewise created another important change in federal government policy in providing for coöperative fire protection with the states carrying on protective work. This section of the Weeks Law launched the federal government upon a coöperative policy with the states, a principle which is destined to play a most important rôle in American forest management.

This epochal piece of legislation authorized the expenditure during the succeeding five years of $11,000,000 for the examination, survey, and acquisition of forest lands in the East. In 1916 $3,000,000 more was added. Similar appropriations followed these. The forest-protection coöperation was provided on condition that any state receiving federal aid must have a law providing for forest-fire protection and must spend at least as much as the federal government in such fire protective work.

A few months after the passage of this law, sale proposals for over a million and a quarter acres had been received. Under this and its amendatory laws up to 1942 more than 18,000,000 acres have been acquired and additional areas have been approved for purchase. Practically all this land is in the eastern half of the United States. In 1911, the first year of coöperative fire protection, eleven states entered into agreement to protect a total of 7,000,000 acres at a total cost of $350,000. For the fiscal year 1938 federal allotments were made to thirty-nine states for about $1,500,000, protecting over 252,000,000 acres of forest land.

Other Federal Land Problems

With the passage of the Weeks Law and its amendatory acts, the National Forest policy, both in the West and in the East, may be said to be on its way to full fruition. As already indicated, the forestry movement at first met with much local opposition, and bitter fights were waged, especially in the West, to defeat the whole enterprise. But gradually, through an extremely well-conceived administrative policy, opposition changed to strong support. Local administrative officers, with powers of local supervision, were hired, and the forests were made to serve local needs, often without cost to the user. Furthermore, forest personnel were always ready to discuss and work out local problems with the members of the industries concerned.

The loss of local tax money through government ownership was more or less offset by the annual payment of 25 per cent of the National Forest gross receipts to the counties in which the forests lie, in lieu of these taxes, to be spent for schools and roads. The further heavy expenditures in and near the forests by the Forest Service for roads, trails, telephone lines and other improvements caused the public to look upon Uncle Sam as a highly desirable neighbor and land owner. Local people saw that all these improvements meant the opening up and the permanent development of vast forest resources hitherto largely inaccessible.

As a result, instead of demanding the abandonment of the National Forest policy, the demand was for its widespread expansion. Thus in spite of the fact that Congress in 1907 took away from the president the authority (without specific congressional action) to create National Forests in six western states, and later in three others, many additional areas have been added upon the demand of local residents. Such lands were often grazing lands and those valuable for watershed protection.

Since the management of National Forests is complicated in many ways by the interdispersal of private lands within their boundaries, a number of efforts have been made to help eliminate these private holdings. Through local initiative and effort, Congress passed a number of acts authorizing the Forest Service to acquire privately owned lands within certain areas by the exchange therefor of an equal value of National Forest land or timber. In 1922 Congress passed the general exchange law authorizing this practice on all National Forests. At first this was operative only within the Forests, but local interests have asked that the principle be applied to areas outside National Forest boundaries. Up to the end of 1931 over 1,200,000 acres have been so acquired. This act will have great weight in consolidating National Forest lands which in turn will result in more efficient administration.

An effort has been made to place under Forest Service management the large areas of Public Domain, already mentioned, which are not now under any form of management. The Forest Service in 1931 proposed to President Hoover's Committee on Conservation and Administration of the Public Domain, certain additions to existing National Forests and certain new forest units suitable for management similar to that of the National Forests. This area consists of about 35,000,000 acres of unreserved public domain, state lands, and private lands. The exchange acts could again be applied to some of these areas. The Taylor Grazing Control Act of 1934 has placed 139,000,000 acres of the unreserved public domain, suitable for grazing purposes, under administration. Much still remains to be done.

One can see that the acquisition of lands for National Forests is gaining ever-increasing momentum. A score of eastern states, by legislative action, have asked to have the federal government establish National Forests within their borders under the Weeks Law. The purchase program under this act and its amendments has been extended to the cut-over lands in the Lake states, southern states, and the western states. Moreover, several private owners have expressed the desire to deed their cut-over areas to the federal government to be added to the National Forests.

Thus it is very apparent that the public is thoroughly sold on the idea of government-owned forest resources, and that they evidently feel that the National Forest policy will mean social, economic, and industrial stability to the forested states. The states have thoroughly endorsed the policy of "the greatest good to the greatest number of people in the long run."

The Clarke-McNary Act

Previous to 1924 the government conceived its duty to be of a four-fold nature: to create National Forests on unreserved public lands; to administer such forests for the common good; to purchase land for National Forests at the headwaters of navigable streams; and to coöperate in fire protection with the states.

Up to this time there had been no basic statute upon which to build a future American forest policy for continuous timber production. There had been no purchase of forest land primarily for timber production. There was no coöperation with the states in reforestation or in forest management. There was no large-scale authority for coöperation with private individuals or corporations in forestry matters. No attempt had been made to create into National Forests the large areas of forest land still in the unreserved public domain. With about 75 per cent of the forest land of the country still in private hands, it would seem that a national plan for American forestry was still in the embryonic stages of development.

The Chief Foresters upon whom devolved the task of bringing into being such fundamental legislation are H. S. Graves (1910-20), William B. Greeley (1920-28), Robert Y. Stuart (1928-33), F. A. Silcox (1933-39), and E. H. Clapp (1939-43). They are largely responsible for the basic legislation that was enacted and their labors were performed often at great personal sacrifice. Stuart and Silcox died while in office. The present Chief Forester is Lyle F. Watts.

During the World War (1914-18) no significant legislation was passed, but under the surface public sentiment was being crystallized, extensive hearings and investigations were being made on the general forest situation, and Congressional support was being mobilized. As a result Congress passed and President Coolidge (1923-29) signed three bills of the utmost importance to American forestry.

The first of these was the Clarke-McNary Act, signed June 7, 1924, which established a more or less definite American forest policy by statute. A special committee of the U.S. Senate under the chairmanship of Senator McNary of Oregon held extended hearings throughout the country with public officials, lumbermen, and foresters. The act gives a statutory basis for coöperative fire protection and suppression on government, state, and private lands, and it is the first law to emphasize the continuous production of timber.

The law states that in no case shall the amount expended by the federal government exceed the amounts spent by the coöperative agencies, and that the coöperative protection shall extend to watersheds of navigable streams, timbered lands, forest-producing lands, and watersheds from which water is secured for domestic use or irrigation. This is a distinct advance over the Weeks Law which provided for coöperative protection only on the watersheds of navigable streams and did not provide for coöperation with private individuals through the states. Another section of the act provides for coöperation with the states on a similar basis for the investigation of forest taxation and forest insurance, with the avowed purpose of encouraging the growing of timber. Another section provides for the production and distribution of nursery stock for establishing windbreaks, shelterbelts, and farm woodlots. Another section is devised to assist owners of farms in establishing, improving, and renewing forest growth.

Other sections of the law are non-coöperative in character. The Weeks Law is amended so that the purchase of new forest lands can be extended to lands valuable for the production of timber. The federal government is permitted to accept forest lands donated by owners. Finally, the law gives authority to determine the location of public lands chiefly valuable for stream flow protection or for timber production which could economically be administered as National Forests. This section aims to incorporate the forested areas of the unreserved public domain into present Forest units or new ones.

While there are other desirable features in the Clarke-McNary Act, certainly the one feature that stands out is the great impetus it gives to federal aid to states. This policy may be said by this act to have been definitely incorporated into the broad American forestry program of the future. There is also a wise and significant clause which states that if the system and practice of the states substantially promotes the fundamental objects of the act, the money will be forthcoming. Otherwise, we suspect, it will not be. It appears that this paves the way for coöperation wherein the federal government will in effect have some say as to the conditions under which the money will be granted. In this way it will encourage state forestry, improve state forestry practices, and bring about uniformity in state methods.

Before leaving the discussion of this important act, it would be quite natural to inquire what the net results have been, even in the face of inadequate appropriations. Fire protection standards have been raised in many states to meet federal standards. Federal coöperation has been instrumental in establishing twelve new State

Forestry Departments and the protection of private forest land was commenced as a state activity in more than seventeen states. Also under federal coöperation the area of state and private forest land receiving organized fire protection was increased from 95,000,000 acres in 1915 to over 290,000,000 in 1942. During the same period federal, state, and private funds devoted to forest fire protection increased many fold. On protected areas forest fires have annually burned over about 1.7 per cent of the area protected, whereas fires have swept 19.8 per cent of the unprotected areas. This certainly speaks for the efficiency of the coöperative system. However, only about 75 per cent of the state and private forest lands classed as in need of protection was actually being protected in 1942.

During the fiscal year 1933 some forty states engaged in forest planting with federal aid, and of these, eighteen owe their start in this extension work to government assistance. From 1926 to 1931, under federal coöperative projects, a total of 156,000,000 trees covering some 150,000 acres were distributed to farmers. In farm forestry extension work a total of over 10,000 farms adopted improved forestry methods by the year 1926, and the number of acres was more than doubled by 1929.

The McSweeney-McNary and Other Acts

What the Clarke-McNary Act did for coöperative forestry work, the McSweeney-McNary Act accomplished for forest research. The latter act, signed by President Coolidge on March 22, 1928, puts forest research on a statutory basis and makes it a definite part of the government forest policy. The bill outlined a research program including investigations in all the major branches of forestry and many allied fields and authorized the Secretary of Agriculture to get the research facts not only from his own Bureaus but also coöperatively from state, private, and even foreign sources. The machinery set up for the work consists largely of twelve forest experiment stations and a forest products laboratory. The ten-year financial program authorized the expenditure of over $30,000,000, although actual appropriations failed to reach that figure. Lastly, it provided for an inventory of forest land and timber, an appraisal of present and potential forest growth, and the data needed to secure the balancing of the national timber budget. While the ten-year program ran to 1938, the bill authorized such annual appropriations thereafter as may be necessary to carry out the provisions of the act.

On April 30, 1928, President Coolidge affixed his name to the so-called McNary-Woodruff Bill authorizing an additional $8,000,000 for the purchase of forest lands to be made into National Forests.

The most recent constructive pieces of legislation are contained in the Fulmer Bill (1935) * which provides for federal coöperation in the purchase of state forest lands, and the Norris-Doxey Coöperative Farm Forestry Law (1937).

The Copeland Report

As the result of a Senate Resolution introduced by Senator Royal S. Copeland of New York in the 72nd Congress, first session, 1932, a voluminous report of some 1,700 pages was compiled by the Forest Service, entitled "A National Plan for American Forestry." This was published in 1933 and is the most comprehensive and exhaustive survey ever made of the forest situation in the United States. Its length precludes any full outline of its contents, but a few of the major conclusions must be mentioned here.

The Copeland report pointed out that the forest problem, with its many ramifications, is one of the greatest problems the country has ever faced; that practically all the vexing problems of American forestry have grown out of the abuses of private ownership; and that in spite of the millions that were being poured into forestry in the United States, the main problem—that of the management of the forest lands under private ownership—remained practically untouched.

The report was epoch-making for American forestry, because it demonstrated clearly the great importance of solving the problems arising from private ownership. Ninety per cent of the total area devastated and poorly stocked, and 95 per cent of the current devastation was on privately owned timberland. Practically all forest deterioration caused by fire or cutting was taking place on private land. The reversion of cut-over land to states for unpaid taxes in several forest regions had led to a practical breakdown in private ownership, to say nothing of its effect on local social, economic, and industrial life. It was impossible to make even a rough guess as to when our national timber budget would be balanced—that is, when the amount of annual growth would equal the annual cut—because 90 per cent of the potential growing capacity of our forest lands was in private control. Floods, erosion, grazing, and wild-life problems baffled solution for the same reason. Finally, the absurd situation whereby a mere 5 per cent of the current forestry expenditures (both public and private) was being diffused over this same 90 per cent of

* Inoperative, since no appropriations have been made by Congress.

our potential growing capacity was exceedingly wretched national economy to say the least.

Finally, the report recommended among other things the immediate purchase of 224,000,000 acres of forest land, 90,000,000 by the states and 134,000,000 by the federal government; greatly increased federal aid to states; more intensive management of public forest lands; the federal adminitsration of the remnants of the forested public domain; and the management for forestry purposes of over 50,000,000 acres of abandoned "marginal" agricultural lands.

Joint Congressional Committee on Forestry

The President of the United States on March 14, 1938, called the attention of Congress to the nation's forest problem and recommended a study by a joint congressional committee. In June of the same year Congress approved the resolution and established a joint committee of five Senators and five members of the House who were directed to hold hearings and to investigate and report on the general forest situation.

The joint congressional committee held public hearings in the major forest regions as well as in Washington, D. C. They gathered testimony covering a wide range of interests—the general public, labor, forest and allied industries, industrialists, small forest owners, and educational authorities. The final report entitled *Forest Lands of the United States,* of 2,163 pages, was published in 1941.

The recommendations of this committee were as follows:

1. The extension of coöperative forest fire protection under the Clarke-McNary Act, together with some measure of public regulation of cutting and related forest practices on private lands.

2. The extension of coöperative protection against forest insects and diseases under the Clarke-McNary Act.

3. The extension of the Clarke-McNary Act to authorize the furnishing of tree seed and seedlings to all landowners.

4. The expansion of the provisions of the Norris-Doxey Coöperative Farm Forestry Act dealing with forest extension work.

5. Legislation authorizing the establishment of coöperative sustained-yield units to allow sustained-yield management of mixed public and private holdings.

6. Provision for a forest credit system to make long-term, low-interest rate loans to private forest and naval stores operators through facilities already available in such agencies as the Reconstruction Finance Corporation, the Farm Credit Administration, and the Federal Land Bank.

7. Legislation authorizing coöperation with the states in encouragement and development of farm forest coöperatives.

8. Legislation providing for leases and coöperative agreements with private forest land owners, communities, institutions, and states.

9. Extension of forest products research and forest management research at the Forest Products Laboratory and the forest experiment stations, with special efforts to get the results of research to the public.

10. Extension of the State Forest Acquisition Act of 1935 and the authorizing of increased appropriations.

11. Increase of federal forest land acquisition.

12. Legislation authorizing financial contributions to local governments in lieu of taxes on forest land removed from the tax rolls by federal acquisition.

13. More adequate protection against fires, insects, and diseases on the National Forests.

14. More intensified National Forest management.

15. An annual appropriation of $750,000 for the completion of the forest survey.

16. Investigation of the apparent monopolistic purchasing of pulpwood by pulp and paper mills.

These recommendations were embodied in the Bankhead Omnibus Bill which Congress failed to pass. Later some of the specific recommendations were included in individual bills. The president signed a bill authorizing coöperative sustained-yield units in March, 1944; another one providing additional funds for coöperative forest fire protection under the Clarke-McNary Act in May, 1944; and one providing additional forest survey funds during the same month.

Recent Federal Emergency Conservation Work

The duty of a government to care for its unfortunate citizens in times of severe economic depression, when other agencies usually fail, scarcely admits of argument. Economic depressions continue to come and go and the problem seems to be not so much how to get rid of them as how to live with them. Our experience since 1933 has shown that forestry work can absorb a great share of unemployment in times of emergency and that a great deal of necessary forestry work can be done which is apparently impossible in prosperous times.

President Franklin D. Roosevelt was evidently very much in sympathy with this method of giving employment and relief, for immediately after his inauguration on March 4, 1933, there followed a series of presidential orders and congressional acts which not only supplied jobs so desperately needed in the depression, but which also accomplished highly important forestry, erosion, and flood control work. Moreover, this work took hundreds of thousands of idle American youths off the streets and gave them healthful outdoor work. In so doing, the president gave a great impetus to the forestry movement and set an example that other presidents might well follow.

The newly created federal organizations which were concerned directly or indirectly with conservation work were the Federal Emergency Conservation Work (E.C.W.) which included the Civilian Conservation Corps, the Public Works Administration (P.W.A.), the Tennessee Valley Authority (T.V.A.), the Soil Conservation Service, the Civil Works Administration (C.W.A.), the Federal Emergency Relief Administration (F.E.R.A.), and a number of others. There were a number of different methods by which money was made available to these organizations. In some cases it was by direct congressional appropriation, either through existing organizations or through new set-ups. In some instances it was by executive order from existing emergency funds held by the president. Some were made as loans or grants in lieu of regular appropriations.

Emergency conservation work inaugurated under this program included a vast number of separate projects, the most important of which were:

1. The Civilian Conservation Corps, assigned mainly to forest work.

2. The Tennessee Valley Authority, which has a basic forestry program.

3. The Prairie States Forestry Project.

4. An act to relieve unemployment which provided the basis for fire prevention and suppression, flood control, reforestation, and other forestry work.

5. The increased purchase of forest land for National Forests and of submarginal farm lands for the same purpose.

6. Wild-life sanctuaries provided in National Forests.

7. Increased coöperation with the states and with private owners of forest land.

The Civilian Conservation Corps was set up by President Roosevelt as a relief measure in April, 1933. Its purpose was to take care of boys for whom there was no place in school or in industry, and the camps were established for the major purpose of promoting their welfare and improving their training. It grew into a major program dealing with relief and conservation. Its main forestry objective was to help protect, develop, and perpetuate existing forests. To carry out this objective it helped to employ some 1,500 or more foresters who were out of work in 1933. The Corps was abolished by Congress in June, 1942.

In August, 1940, the 1,500 camps were filled with about 300,000 enrollees from every state in the Union. During nine years over 3,500,000 men were enrolled in the work, and the direct accomplishments can be summarized as follows:

1. About 2,355,000,000 young trees planted.
2. More than 126,000 miles of roads and trails built.
3. 43,000 bridges constructed.
4. 78,000 miles of telephone lines built.
5. 6,500,000 man-days of fire fighting.
6. 5,200,000 erosion control check dams built.
7. Over 4,000,000 acres of forest growth improved to increase the growth.

Of course, there was much other important work done, not to mention the discipline and training secured and the mental, moral, social, and spiritual values that have been acquired by the men in the work. Finally, over $600,000,000 was sent home by these enrollees to their needy dependents.

The Act creating the Tennessee Valley Authority was approved by the President on May 18, 1933. This organization is actually a corporation with the powers of government but with the flexibility of a private enterprise. It was charged with planning for the proper use, conservation, and development of the natural resources of the Tennessee River drainage basin. This involves an area of some 40,000 square miles in parts of seven states and with a population of some 2,000,000—with three times that number living in the adjoining territory. The Authority actually grew out of the Muscle Shoals power development which had been more or less idle since the First World War. This, together wtih the Wheeler and Norris power developments on the same river, represents an investment of over $110,000,000.

The Tennessee Valley covers some 28,000,000 acres, of which 63 per cent is forest land, nearly all of which had greatly deteriorated. About 800,000 acres have been eroded, some being in a critical condition. There are thousands of acres of abandoned farm lands that should never have been cleared because of their steepness. The forestry work involves forest fire control, soil erosion control, timber stand improvement, reforestation, installation of forest nurseries, forest surveys, the building of roads and trails, and forest education work. All this has a direct bearing upon the permanence of the investment made in water power development. In fact, T.V.A. is a gigantic object lesson in protective forestry which may have a far-reaching effect upon our future forest policy with respect to the rôle of forests in preventing floods and erosion.

The Prairie States Forestry Project (first organized as the Plains Shelterbelt Project) proposed a series of shelterbelts from North Dakota to Texas for the purpose of preventing wind erosion and excessive drying of the soil. Plantings up to June, 1942, totaled over 222,800,000 trees in over 18,000 miles of field windbreaks on over 33,000 farms. Examinations of five years of planting activity show an average survival of well over 65 per cent.

During 1933 and 1934 President Roosevelt allotted $30,000,000 for the acquisition of additional National Forest lands from the Emergency Conservation Fund. Special funds were also made available for fighting the Dutch elm disease and the white pine blister rust. P.W.A. allotted over $30,000,000 for National Forest improvements, and the same administration allotted $20,000,000 to the Soil Conservation Service for erosion control projects.

Summary of Federal Forestry and Related Activities

The outstanding contribution to federal forestry to date has been the creation, administration, and protection of the National Forests which today cover over 176,000,000 acres. The organization of the Forest Service which has charge of these Forests was lately modified so that there are five major branches: Administrative Management and Information; National Forest Administration; Land Acquisition and Land Planning; State and Private Forestry; and Research. In the field there are forest and range experiment stations, a Forest Products Laboratory, and many other specialized projects.

The field organization is set up on a geographical basis to expedite forestry business. Ten regions cover the country and each has a regional headquarters in some city centrally located, in charge of a Regional Forester and staff. The executive, administrative, and protective units are the National Forests covering on the average a little over one million acres each. The 160 Forests are divided into some 800 ranger districts. The total personnel for the last few years has been well over 6,000. The entire force is under Civil Service regulations and it is no exaggeration to say that it is one of the most efficient and faithful government services we have. This is largely due to the long uninterrupted service of its personnel and to the comparative freedom from political patronage.

The intelligent management of the nation's forests demands reliable information upon their resources and the conditions under which they may be best perpetuated. This research covers most branches of forestry work, under a great variety of conditions. It involves problems in growing and protecting timber; in developing grazing and wild-life values; in better watershed management; in better utilization of forest products; and in hundreds of other activities. For this purpose twelve forest and range experiment stations, located in the various forest regions of the country, have been set up.

The Forest Products Laboratory at Madison, Wisconsin, is part of this research organization. The work of the laboratory is directed towards developing methods to reduce waste and to lower costs in the logging, manufacture, and utilization of wood products; to increase the serviceability of forest products to the user; and to create new and useful products from wood. The laboratory is provided with plant facilities and experimental equipment designed for the handling, processing, .testing, and investigation of wood in many forms from the raw material to lumber, pulp, turpentine, plastics, and other conversion products.

A nation-wide forest survey was authorized by the McSweeney-McNary Act of 1928. It was started in 1930, and about one-half of our forest land, or 300,000,000 acres, has been covered. The total cost will be around $6,000,000. Its object was to make an inventory of our timber supply; to ascertain the rate of growth; to ascertain the rate at which the supply is being diminished by actual use, fire, and disease; to determine probable future requirements of timber and forest products; and to use the results to formulate a better national timber economy.

The Soil Erosion Service, in the Department of the Interior in 1933, was transferred in March, 1935, to the Department of Agriculture, and was later called the Soil Conservation Service. Since then it has been enlarged to include many land-use projects, especially those under the Bankhead-Jones Farm Tenant Act of 1937, the Coöperative Farm Forestry Act of 1937, and what was left of the Prairie States Forestry Program. It coöperates with other bureaus of the Department in the many varied problems of better land use.

The service activities relating to forestry cover such phases as: a nation-wide network of erosion-control demonstration projects; soil conservation practices on 793 soil conservation districts established in 41 states; land acquisition; coöperative farm forestry; conservation nurseries; flood control; and research. The fundamental aim of these activities is to make each acre of crop, pasture, range, and forest land yield its maximum of services to the owner.

The Bureau of Plant Industry of the Department of Agriculture is responsible, among other things, for plant disease investigations. Its division of Forest Pathology deals with the diseases of forest and shade trees and the decay and discoloration of timber and other forest products.

The Division of Forest Insect Investigations is one of the several units of the Bureau of Entomology and Plant Quarantine. Its activities pertain to the study of insects affecting forest and shade trees and forest products.

A great deal of forestry work has been carried on by the Department of the Interior for many years, in spite of the fact that the Federal Forests reserved since 1891 were turned over to the Department of Agriculture in 1905. The forestry work of the Department of the Interior is now under the supervision of the Office of Director of Forests, created in 1939. Most of the forestry work comes under the following main divisions:

1. Oregon Grant Lands.
2. Forestry on Indian Lands.
3. National Park Service.
4. Interior Forests of Alaska.
5. The Federal Range.
6. The Fish and Wildlife Service.

Due to the non-compliance of the owners of the Oregon and California Railroad and Coos Bay Wagon Road Grants, the federal government instituted proceedings to restore them to government ownership. This was finally accomplished in 1916. It placed in the hands of the Department of the Interior 2,500,000 acres of forest land with a stand of about 50,000,000,000 board feet of virgin timber, mostly Douglas fir. The lands are being managed on a sustained yield basis, and 50 per cent of the income goes to the counties in which they are located, in lieu of taxes.

There are 55,000,000 acres of Indian lands, 42,000,000 of which are suitable for grazing and 16,000,000 of which are covered with forests or woodlands. Of the latter only 6,500,000 acres are of commercial grade. This area contains about 30,000,000,000 board feet of timber.

These Indian lands bring in considerable income. Grazing privileges have been valued at more than a million dollars annually for the past forty years. Since 1910 more than $45,000,000 has been re-

Department of Agriculture Photo

The darker (lined) shaded areas are national forests. The lighter (dotted) areas are purchase units. The larger dots (as at Atlanta) are regional headquarters; the smaller dots, supervisors' headquarters. At Madison, Wis., is the U.S. Forest Products Laboratory.

ceived from timber sales, and it is thought that eventually these lands will sustain an annual cut of 300,000,000 board feet, valued at over $1,000,000 annually. Since these forests are Indian property and not public property, they are being administered and protected like private property by the government acting as trustee.

On June 30, 1941, there were 164 units in the National Park system comprising over 21,000,000 acres. They include many of the most beautiful and remarkable forests, mountain peaks, geysers, hot springs, glaciers, volcanoes, canyons, lakes, and waterfalls in the country, as well as many sites of historic interest.

In 1933 a Branch of Forestry was established in the National Park Service. Its principal duties consist of protecting the Park forests from fire, insects, and forest diseases; mapping the forest types; and conducting studies for the improvement of the forest stands.

The objectives desired in the management of the forests on the National Parks differ radically from those governing the management of other federal forests or of commercial forests. The aim is to preserve them in their primeval state. Timber cutting, grazing, and other commercial uses are forbidden and the lands are devoted primarily to educational, inspirational, and recreational purposes.

The interior forests of Alaska cover an area of from eighty to one hundred million acres of scattered woodlands. They are largely of poor growth and at present inaccessible. Since 1940 an attempt has been made to protect these forests from fire by providing congressional appropriations.

The Taylor Grazing Act of 1934 put an end to the free public range of the West. This Act provided that approximately 140,000,000 of the 185,000,000 acres of unappropriated and unreserved public lands be administered for the benefit of the range resources and the advancement of the livestock industry. This grazing land is administered in ten grazing districts with general headquarters in Salt Lake City. On these grazing lands and also on the still vacant, unappropriated, and unreserved public domain there remain a total of about 25,000,000 acres of forests and woodlands, and at present there exists no legal authority for their management, even though they are under the jurisdiction of the Department of the Interior.

The Fish and Wildlife Service established in June, 1940, consolidated the work formerly done by the Bureau of Biological Survey and the Bureau of Fisheries. The work deals with game and other birds; game, fur, and other mammals; reptiles and amphibians; commercial and game fisheries; shellfish industries; and fur seals, whales, and marine mammals.

The Service administrates several federal statutes for the protection and conservation of migratory game and other birds; of game, fur, and other mammals; and of fish. It also establishes, administrates, and operates game refuges. Up to June 30, 1941, there had been created 267 national wild-life refuges totalling over 13,740,000 acres. Many of these wild-life areas are stocked with forests, and the foresters of the Service develop these stands so that they will be of the greatest use as producers of food and shelter, and as habitats for wild life. The Service also carries on an extended program of research and enforces the conservation laws relating to wild life and fisheries that come under its jurisdiction.

No picture of federal forestry would be complete without a warm acknowledgment of the persevering labors of numerous private organizations, each of which has carried its little contribution of "bricks" and "mortar" to rear the forestry edifice of the nation. In addition to the organizations already mentioned elsewhere, we should note the American Tree Association, American Nature Association, the Camp Fire Club of America, the Ecological Society of America, the National Parks Association, American Planning and Civic Association, Save the Redwoods League, and the Wilderness Society.

CHAPTER XV

PROGRESS OF THE STATES IN FOREST RESTORATION

MILLIONS OF ACRES of land unsuited to agricultural use and once covered with dense stands of timber are today unproductive wastes, a liability to the states in which they are found. While the federal government has taken over some of this land by purchase, many areas are distinctly adapted for state ownership. The National Resources Committee has recommended that state ownership be eventually expanded to include 77,000,000 acres, mostly in the densely populated states where the public will reap the full benefits through timber production, watershed protection, erosion control, recreation, and improved wild-life conditions. The organization, protection, and improvement of these forests will be the work of the state forestry departments for many decades to come.

One of the activities of the various state forestry departments is the acquisition and administration of acquired forest lands.* Thirty-nine states now own a total of approximately 13,400,000 acres of state forests in 732 units. States ranking highest in this respect are:

Minnesota	5,338,238	acres
New York	2,674,473	"
Pennsylvania	1,650,937	"
Michigan	1,000,000	"
Montana	520,000	"
Idaho	448,000	"
Washington	355,000	"
New Mexico	258,000	"
Massachusetts	171,360	"
Wisconsin	170,190	"

In contrast with the large forest areas in these states, thirteen southern states combined have only 143,833 acres of state forests.

* *State Forests for Public Use,* U.S. Department of Agriculture, Forest Service, Misc. Pub. 373, 1940.

State appropriations for the year 1937-8 totalled about $7,240,000. This money was spent for administration, protection, reforestation, purchase of forest land, maintenance and improvement of state forests, research, education, and extension work.

Expenditures for the protection of state and private forest lands against forest fires during the year 1937 were over $7,600,000, of which about $1,500,000 was federal money alloted to the states under the Clarke-McNary Law.* However, it is important to note that only 75 per cent of the total area in need of protection is now being covered, and that coverage is not intensive enough to prevent heavy losses. Some 136,000,000 acres of woodland are without protection, and of this about 20 per cent burns over every year at a loss of $30,000,000 to $40,000,000. Experience shows that the total forest fire bill will have to be about $18,500,000 if all the forest land (outside of National Forests and Parks) is to be adequately protected. Needless to say our present meagre expenditures are poor economy.

A most important function of our present and future state forests is to provide recreational facilities for the millions in the great population centers. The latest available figures show that about 28,000,000 people visit the state forests in a single year. Many types of facilities are offered. Camp grounds, picnic areas, trailer camps, and summer cottages are usually available. Public camp sites usually have rough fireplaces, tables, benches, pure water supplies, and sanitation equipment. Some have bathing facilities. In the winter, appropriate facilities are often available for tobogganing, skiing, skating, and ice boating.

Another important state forest activity is reforestation. In the remoter parts of the forests, away from the recreational areas, forests are being restored for timber production. Over 670,000,000 trees have already been planted on state forests, not only to provide for a future timber crop but also to control floods, reduce erosion, provide food and cover for wild life, and to reclaim abandoned farm lands suited for forests. Trees are also being furnished to private landowners at cost, and federal assistance through the Clarke-McNary Act has been given the states for producing trees supplied to farmers.

The state forests are also helping the hunters, for they are providing food, shelter, and breeding places for quail, partridge, pheasants, rabbits, squirrels, deer, bears, and other types of forest game. State forestry departments are coöperating with the state game depart-

* In 1940 under this act the federal government allotted $2,154,444 for coöperative fire protection; the states' share was $5,154,941, and private owners contributed $2,045,243.

ments to protect game, to raise young game animals, to provide food by planting trees, shrubs, and other plants, to conduct game censuses, to establish game refuges, and to conduct studies on the best habitat conditions for various types of game.

Fishermen are looking to the state forestry officials for improved fishing conditions. The maintenance and protection of a good forest cover are essential to clear and cold streams that harbor trout, salmon, and other game fish. Hundreds of miles of improved trout streams have been developed by building dams, by increasing the aeration of trout waters, and by checking erosion along stream banks by building cribbing. Hundreds of millions of game fish of various kinds and sizes have been raised every year by state fish and game departments, as part of the broad state conservation program.

The Burden of Responsibility Resting Upon the States

No one can doubt the great importance of federal forestry work on its National Forests and in its coöperative acts. It must be remembered, however, that at the present time less than 25 per cent of our forest lands are under the control of the national government. It is thus inescapable that for many years to come (regardless of possible large federal acquisitions) the heaviest burden of responsibility for initiating and developing forestry upon the remaining 75 per cent falls upon the individual states, and through them upon the private owner.

What the private owner can do in the way of practising forestry will depend directly upon the state's attitude towards taxation, protection, reforestation, regulation of cut, and other matters. The private owner is entirely at the mercy of the state and local governments in these important matters. Even should the federal government double its present holdings of commercial forests, the state would still be directly or indirectly a most powerful factor in establishing American forest policy.

To put it another way: as long as three-fourths of our forest lands are in private hands with no prospect of management for continuity, our American forestry problem is only about one-fourth solved. How far can we go in public regulation? Can we force better forest practices upon the private owner through mandatory legislation? Is public regulation of private forests an undemocratic invasion of our Bill of Rights? These are questions not easily answered.

We have just passed through a decade of great social advances and increased government control. Can this be applied to forestry also? Social control may be used effectively, perhaps, to stop gross abuses of private property, but its use to enforce forestry practice of private owners, where economic conditions are not ready for it, is extremely questionable. The best that the government can do now is to create, as far as it has the power, favorable economic conditions and to supply up-to-date information through intensive research. The private owner not only needs to know more about growing forest crops, but also more about cutting systems, logging methods, manufacturing and marketing, so as to reduce costs and increase returns. Private owners will fall into line when they are convinced that it will pay them to do so. But the attitude of the federal and the state governments will largely determine how rapidly this will come about.

While the rise or fall of American forestry depends on what happens to the bulk of privately owned forest lands, the separate states have their own spheres of activity in which they must remain supreme. Aside from the task outlined above, it must be the duty of the states to acquire up to a total of about 75,000,000 acres of state forest lands. These probably will be made up largely of watershed forests, tax-abandoned cut-over lands, and abandoned, sub-marginal farm lands. In the main they should be the less profitable forest lands which might be unattractive to the private investor. Where states are financially unable to carry out this program, it may have to be done with federal aid.

The state should also be the middleman of coöperation between the federal government and the private forest land owner in matters of fire protection, reforestation, forest management, the marketing of forest products, and in other spheres of activity. States must also carry a considerable load of forest research, improve forest taxation laws to promote forestry, support state forestry instruction, and produce and distribute forest planting stock. It must enact statutes to permit the acquisition of forest lands by counties, towns, and cities. Eventually, the state may be compelled to pass regulatory laws to induce the private owner to protect watershed forests and even to keep his forest lands continually productive. As we shall see, many states have made a beginning in some of these activities, and a few states have made progress in all of them.

Diversity of State Accomplishment

In surveying the forestry achievements of the forty-eight separate states, one is struck by the great diversity of accomplishment. It is evident that a vast variety of social, industrial, economic, climatic, and physiographic conditions have affected the creation and the development of state forestry activities in each region of the country and also in the different states of a given region.

Each state did not develop its forest policy independently. The first wave of enthusiasm started after the Civil War. Most of the states were caught in the general agitation for forestry reform which began to take shape in the first decade of the twentieth century. This was nation-wide and federal in character. The federal government set the pace first by force of example and later by coöperative aid. The laws of 1911 and 1924, by holding out the promise of federal funds, brought a great many more states into line, particularly the southern. At the present time only Arizona, New Mexico, South Dakota, and Wyoming have no organized state departments; while Nevada and Nebraska organize the work under extension foresters.

The amount of accomplishment is not necessarily in proportion to the area of state forests. The northeastern states, the Lake states, and the northwestern states have done the best job in that order. The Prairie states for the most part are treeless, and the Rocky Mountain states and the Pacific Coast states have large areas of federal forests. Arizona and New Mexico handle their state lands by coöperative agreement with the U.S. Forest Service.

The principal factor which seems to determine the speed of development of state forestry is the percentage of state land covered with forests and the degree to which this forest wealth has been exhausted. Large areas of cut-over land that needs planting, uncontrolled forest fires, large importations of lumber and forest products from other states—these and other conditions lead to a public awakening to the value and need of forestry. Hence, the northeastern states, which have had such conditions for a long time, started early and now lead in forestry achievement.

Maine, New Hampshire, and Vermont are largely rural and have a large percentage of their area in forest. In fact, the first two of these states are over 75 per cent forested. Hence, forest fire protection is regarded primarily as a public function. As might be expected, state forests have been created. New Hampshire has about 42,000 acres and Vermont about 50,000 acres under administration. Further acquisition is hampered by financial circumstances. Southern New England is highly industrialized. The consequent larger tax base enables these states to finance their forestry needs more adequately. Massachusetts has 170,000 acres of state forests and Connecticut over 70,000 acres under administration. Fire suppression in these states is done largely through local organizations.

Industrial development means a need for lumber, paper, and

other forest by-products. Dense population near industrial centers demands recreational facilities. It takes more than fire-scarred hill-sides to furnish fish, game, and restful camping. Wealthy states with basic industries and large natural resources have the taxing capacity which is necessary to provide large state appropriations. Hence the great strides taken by such states as New York, New Jersey, and Pennsylvania. Of these the last named has the best balanced program.

The three states just mentioned have approximately 40 per cent of their areas covered with forests. The acquisition and administration of state forests is therefore paramount. New York now has about 2,700,000 acres, Pennsylvania over 1,600,000 acres, and New Jersey over 55,000 acres under administration. All these states have highly efficient, centralized, fire-protective systems. As might be expected, with much cut-over land, reforestation activities have reached a high stage of development, especially in New York and Pennsylvania.

Michigan, Minnesota, and Wisconsin have from 40 to 50 per cent of their land areas in forests. The era of exploitation was somewhat later in these states than in the Northeast, hence forestry got a later start. The two outstanding problems are fire protection and the utilization of the vast areas of non-agricultural cut-over lands. State acquisition is therefore important. Michigan has over 1,000,000 acres of state forests and Minnesota over 5,000,000 acres, while Wisconsin has 170,000 acres. The administration of these forests is still far behind that encountered in the northeastern states. Large additions to state forests will no doubt be made through tax forfeiture. Administration, including reforestation, will, however, lag because these states are handicapped financially by a small tax base.

The nature of forest distribution—whether it is in the form of small woodlots or in large commercial tracts—may greatly affect the amount and kind of forestry work. Woodlot problems are peculiar to such states as Ohio, Indiana, and Illinois. Forestry is important in these central states because forest products are one of the major resources of their farms.

The southern states were the last to feel the economic results of timber exhaustion. Hence most of the growth of state forestry activities has taken place in the last ten to fifteen years. Virginia, North Carolina, Louisiana, and Texas were the first to organize. The co-operative features of the Clarke-McNary Act of 1924 gave forestry in the South a much-needed impetus. Now every southern state has a forestry department. In these states the principal activity is fire protection. There are comparatively few state forests. The field of forestry activity, especially reforestation, is as fertile as in the Lake states, but the financial condition of the states is such that much of the work will have to be done in coöperation with the federal government.

Since the National Forests of the western states embrace more than one-half the commercial forest land of the region, state forestry in these states has been eclipsed by federal activity. State forestry policies, however, have been strongly influenced by the example set by the federal government. The outstanding problem in this region is fire protection. On account of the complexity of ownership, activity must be coöperative, with private, state, and federal lands bearing the burden proportionately. State forest land administration has grown out of the early federal grants of "school" land which have been more or less consolidated by exchange with federal and other lands. Montana, Idaho, Washington, and Oregon have some 1,400,000 acres of state forests. In the Pacific Coast and northwestern states the problem of cut-over lands is rapidly coming to the front. Large additions to state forests doubtless will be made in the future through tax forfeiture.

Progress of Individual States

The development of state forestry in many ways parallels that of federal forestry. Since the latter has already been detailed in a previous chapter, only a few high lights will be enumerated here. It will be recalled that early colonial ordinances were designed to safeguard the future timber supply by preventing trespass and forest fires. From that time until 1885 state forestry got off to a number of false starts, due to misconceptions about the idea of forestry.

The idea that forestry was synonymous with "timber culture" and "tree planting" gave rise, beginning about 1860, to a host of laws by various states attempting to advance forestry by offering tax exemptions or bounties for planting land to trees. Very little of permanent value was accomplished by this legislation so far as a permanent state policy was concerned.

Another movement that had its beginning at about the same time, but with a better promise of success, was the series of laws passed to create forest inquiry commissions. In general their purpose was to inquire into the rate of forest destruction, its effect on climate and stream flow, and what the state should do to preserve its forests. Such commissions in Michigan, Wisconsin, Maine, New York, Connecticut, New Hampshire, Vermont, Ohio, Pennsylvania, and North Carolina between 1867 and 1891 did little more than temporarily focus public attention on the forest problem, although they were in a few cases the forerunners of permanent state forestry departments, as in the case of New York, Wisconsin, Maine, and New Hampshire.

While the first definite federal forestry action came in 1876 and the first forest reserves were created in 1891, the first state efforts fall between these dates. The progressive dissipation of the nation's timber wealth had by this time reached alarming proportions both in the East and in the West. In the East, especially, it was imperative to crystallize public apprehension into a policy of action. The peak of lumbering had passed successively from Maine to New York (1850), to Pennsylvania (1860), and finally to the Lake states in 1892. The lumber industry was casting a wistful eye on the southern pinery. Many western states were deeply solicitous of their forest wealth as the federal land policy was turning this valuable resource over to private interests.

In 1885 the first attempts at state forestry administration are to be noted when Colorado, California, Ohio, and New York created forestry organizations. Only in the last-named state, however, did this initial organization prove to be permanent.

Colorado, as early as 1876, made an attempt to start state forestry when it requested that the public forests be turned over to the state. It depended largely upon these federal forests for irrigation water for its agriculture. The law of 1885 provided for a forest commissioner to make regulations for the protection and conservation of woodlands of the state and to create public interest in forestry. The funds at first appropriated were nominal and they ceased entirely in 1891. In 1897 the office of commissioner was abolished and in its place was created a Department of Forestry, Fish, and Game which was equally unsuccessful. The office of state forester was established in 1911. The State Board of Agriculture, serving as the State Board of Forestry, employs the state forester, who is the Professor of Forestry in the Colorado Agricultural College. Colorado has one state forest and no state nurseries. The state forester is chiefly engaged in education and in extension and investigative work, because the U.S. Forest Service administers about 75 per cent of the forest area of the state.

California started off with a vigorous interest in state forestry in 1885. The State Board of Forestry has the honor of being the first agency created by any state to have charge of forestry matters. The duties of the board were largely informational and investigative, but forest planting and the protection of water resources also received encouragement. Later law enforcement and protection against forest fires were added.

California also made an attempt to get control of the federal forest lands within its borders in 1887. The agricultural interests of the state were greatly disturbed at seeing their great Sierra forests decimated by pillage and plunder. Federal land law frauds were rampant and most of the best timber passed into private hands. Cutting for the most part was uncontrolled. Sheep men set fire to the slash to improve pasturage. Forest fires ravaged young growth and the fear grew that irrigation, the life of California agriculture, would be irreparably injured. The Sierras were the natural reservoirs for the irrigation of the San Joaquin and the Sacramento valleys.

In spite of a bitter fight to save the forests the agricultural and forestry interests lost. The board was abolished in 1893, and for twelve more years the plundering went on unmolested by either the federal or the state authorities. The establishment of National Forests ended the era of plunder.

Finally in 1905 the State Board of Forestry was created, the first of its kind. This was amended in 1910, and in 1927 a State Department of Natural Resources was created, under which forestry was one of four, later of five divisions. The 1941 annual appropriation by the state for forestry was about $1,268,000, expended mostly for fire protection.

A virile but short-lived forestry movement in Ohio induced the state legislature to create a State Forestry Bureau for the promotion of forestry. Its duties were to ascertain the forest resources of Ohio, investigate causes of forest destruction, and suggest legislation. Meagre appropriations gave rise to little but annual reports. The Bureau ended its career in 1890, due partly to a waning interest but mainly because the time, apparently, was not ripe for the development of forestry. It doubtless had an important educational influence. In 1906 the Act of 1885 was repealed and a Department of Forestry was established at the State Agricultural Experiment Station. In 1913 the work was taken over by the Agricultural Commission but still retained at the station at Wooster. State forests were acquired by purchase under the Act of 1915, and so far over 60,000 acres have been put under administration. In addition there are some 20,000 acres of state parks. The state forester and his staff engage in a variety of forestry work peculiar to the woodlot conditions of the state, viz.: forest protection, maintenance of forest nurseries, forest planting, cooperation with the federal government in fire protection and forest extension work, promoting farm forestry, and investigative work, all of which is now organized under the Division of Forestry.

The early history of forestry in New York State is to some extent a repetition of the pillage and plunder perpetrated on the federal domain. Very early the state was induced to sell its Adirondack forests at ridiculous prices. As early as 1868 the public became interested in stopping this practice and in inaugurating a reservation policy. State lands were bought, stripped of their marketable timber, allowed to revert to the state for unpaid taxes, and then, when the rest of the timber became marketable, would be redeemed from the state, and the stripping of the land continued.

In 1872 the New York legislature appointed a temporary commission to inquire into the matter, but the recommendations were not carried out. It was not until 1883 that another attempt was made by the legislature to put an end to this abuse. The next year it authorized the employment of experts to investigate and report on a system of forest reserves. Dr. Charles S. Sargent of Harvard, a noted botanist and a student of forest conservation, headed this commission. In 1885, as a result of this report, a law written by Dr. Fernow established a permanent forest commission to promote forestry in the state, to supervise the forest preserve, and to organize a system of fire protection. This inaugurated the longest period of continuous operation for any state department administering forest resources, now in existence for over fifty-five years.

This pioneer effort in state forest conservation was many years ahead of any other state or federal legislation. There was no precedent to go by, and the difficulties were many. New York soon stopped the sale of its own timberlands, and by 1883 had accumulated some 700,000 acres of tax-delinquent forest lands. A Civil Service Law was enacted in the same year to introduce the merit system for the appointment of forestry personnel.

The dominant motives in establishing forest reserves were undoubtedly to conserve and regulate water supply and to safeguard recreational and health values in the Adirondack wilderness. There had grown up a strong sentiment among sportsmen and many wealthy and influential people to retain the Adirondack forests in their wild condition, as a summer playground. Hence the cutting of timber on the state lands seeemed in direct opposition to these motives. Though conservationists pointed out that forestry could not be practiced without cutting mature trees, the public and the state press greatly mistrusted their public servants and knew only too well the stealing that had been going on for years. When finally the sale of timber and the leasing of camp sites was permitted, it brought such a storm of protest that when the Constitutional Convention met in 1894 they incorporated in the revision an absolute prohibition of any cutting of timber on state lands or any leasing of such lands. This apparently was good common sense in the light of what had gone before, but it was very poor forestry.

But even the inclusion of this almost irrevocable prohibition in the state constitution by no means stopped the plundering of the state's forest resources. As a matter of fact, for twenty or more years the chief struggle of the state forest department was against political corruption and incompetence. In the face of such conditions progress was slow.

A number of changes have taken place in the name of the commission and the number of its members. It became the Forest, Fish, and Game Commission in 1895 under the direction of five salaried members, reduced in 1901 to three, and in 1903 to one. It became the Conservation Commission in 1911 with three members, again reduced to one in 1915. In 1927, during Governor Smith's reorganization of state departments, the name was changed to Conservation Department under a commissioner and a deputy. There are now five divisions in this Department: Lands and Forests, Fish and Game, Water Power and Control, Parks, and Finance. The Division of Lands and Forests under a director and an assistant has five superintendents to supervise the work: forest fire control, forest pest control, reforestation, camps and trails, and forest investigations. The state is divided into twelve forest districts, each in charge of a district forester.

The reforestation program of New York is unique in the entire nation since it provides for the purchase of abandoned and submarginal farm land for this purpose, at a total outlay of $20,000,000. This measure was ratified as an amendment to the constitution by the people in 1931. Up to the end of 1940 over 400,000 acres have been purchased and over 320,000 have been reforested. In 1936 the Conservation Department set a record in planting over 60,000,000 trees. Since then the number of forest nurseries has been reduced from six to two and the annual output is about 12,000,000 young trees.

A highly developed and efficient forest fire organization covers the forest counties of the state. There are about 100 forest-fire lookout towers and forest-fire observers on state lands. At present 87 rangers are employed. Forest recreation also holds a high place in the state's activities. Over 670,000 individuals used the recreational facilities of the Conservation Department in 1940. New York State's outstanding contributions to state forestry are undoubtedly the administration of 2,700,000 acres of state forests and the record of having planted both on private and on state forest lands about 600,000,000 trees since 1900, when the first state nursery was established.

The experience of still another state is needed to complete our picture of early and successful attempts at forestry organization. This state is Pennsylvania. Although the movement started a little later than in New York, its growth has been steady. Less controversy has occurred and perhaps less political corruption has impeded progress. Although Pennsylvania still has a long way to go, it is nevertheless in the front rank of states in point of accomplishment.

No state in the Union has had its virgin forests more completely wiped out than has the state of Pennsylvania. It was originally almost entirely covered with dense stands of pine, hemlock, and hardwoods. Now these have been cut almost to the last acre. The forest area has been reduced to less than 40 per cent of its original size, and of course what forests remain are almost entirely second growth of poor average quality. Perhaps there is some connection between the cutting of about 30,000,000 acres of dense mountain forests in the last 100 years and the ever-increasing Ohio River floods!

Certainly Pennsylvania is a state in which sound forestry practice is of paramount importance. Its mountainous terrain and its large rivers make watershed protection of vital importance, especially since many of the large industrial cities are in the river valleys where they feel the maximum effects of erosion and floods. The dense population calls for recreational and health facilities and the highly developed industries are intimately bound up with the utilization of wood.

In 1887 Pennsylvania created a temporary commission of inquiry to consider the forestry interests of the state. No legislation was enacted as a result of this inquiry. In 1893 a salaried commission of two was appointed to study forest conditions and make recommendations. As a result, the Act of 1895 created a State Department of Agriculture in which there was a Commissioner of Forestry. His duties were collecting and disseminating information, protecting the forests from fire, and administering state forest lands.

In 1897 the state took two important steps towards acquiring state forests. One act authorized the purchase of forest land at tax sales, while the other created a forest reservation commission to locate and acquire three forest reservations of 40,000 acres each on the watersheds of the Delaware, Susquehanna, and Ohio Rivers. In 1899 the legislature authorized the purchase of wild lands directly from their owners at not over $5 per acre. Thus very soon after forestry legislation began, Pennsylvania realized the importance of acquiring state forests.

So important had forestry become by 1901 that a distinct Department of Forestry was established and the Commissioner of Forestry was made president of the State Forestry Reservation Commission. To this commission was given authority to purchase lands and to manage and control reservations. In 1903 a Deputy Commissioner was provided, and in the same year expenditures for the purchase of forest reserves was limited to $300,000 in any fiscal year.

Pennsylvania made an early start in the education of state foresters. In 1903 an act provided for a school at Mont Alto for the training of forest wardens. This State Forest Academy, the first of its kind in the country, was in 1923 made the Pennsylvania State Forest School. In 1930 the Forest Research Institute was established by the Department of Forests and Waters to provide adequate research facilities. By 1932 Pennsylvania led all states in the number of technical foresters in its employ with a total of 61. All the states together employed only 327, and the eleven southern states, by way of contrast, employed together only 67.

As early as 1905 Pennsylvania provided reimbursement to the townships in which State Forest Reserves were located, inasmuch as these forests were exempt from local taxation. In 1907 the governor approved an act appropriating $500,000 for the purchase of State Forest Reserves.

A very important step was taken in 1909 by an act which authorized the Department of Forestry to grow and distribute young forest trees at cost. Since much of the state land needed reforestation, this act was the beginning of an important reforestation program. Four forest nurseries now have an annual output of about 15,000,000 young trees annually. Up to 1938, 134,000,000 trees had been planted on private lands and 54,000,000 on state lands.

The first attempt at forest fire legislation came in 1897, when constables were made ex-officio fire wardens and their duties were prescribed. In 1909, however, this act was repealed and a new act created a distinct and separate force of fire wardens. At the same time forest fire protection was given added efficiency and stability by making the Commissioner and his Deputy responsible for forest fire protection. The law also authorized the appointment of district fire wardens in each borough and township. In 1915 the legislature established a bureau of forest protection in the State Department of Forestry. There are now 10 districts created for administrative and protective purposes. In 1938 there were 80 forest rangers and 4,400 commissioned forest fire wardens. The state had 145 fire lookout towers, of which number 42 were on state forests.

Laws providing tax-exemptions to promote forestry on private lands were twice (1906 and 1908) declared unconstitutional. Hence the acts of 1913 provided that district foresters should encourage forestry on woodlots and on private forest lands. The acts also provided for the maintenance of private forests as auxiliary state forests which were to be exempt from taxation because of the public benefit derived. Under the Auxiliary Forest Reserve Law land used for forest production is assessed at not over $1 per acre and a yield tax is collected at the time of cutting. Over 45,000 acres of forest land is now so classified.

In 1915 two laws required the counties to offer to the state forest lands taken by them for the non-payment of taxes and also authorized the department to pay as high as $10 per acre for such forest land. It is interesting to note, however, that the 1,650,000 acres so far acquired by the state for state forests was secured at an average of $2.43 per acre.

Although the state laws did not require the head of the Department of Forestry to be a trained forester, Pennsylvania has been fortunate in having two non-technical men of very high calibre in charge of forestry activities from the very beginning to 1920. Dr. Rothrock and Robert F. Conklin, his successor, provided a continuous and vigorous policy for twenty-five years. Since 1920 professional foresters have had charge. The Pennsylvania Forestry Department, now known as the Pennsylvania Department of Forests and Waters, has always had strong public approval and has been fortunate in having governors in sympathy with the work. Gifford Pinchot was the State Forester from 1920 to 1923, after which he was elected to the governorship for two terms, 1923 to 1927 and 1931 to 1935. He will be remembered as one-time Chief Forester of the U.S. Forest Service. With such political and professional backing no state department of forestry has had more stability and more steady and wholesome development.

The recreational program of Pennsylvania is one of the best-developed of all the states in the Union. Over 2,000,000 campers use the state forests annually. There are ten state historical parks, eleven state forest natural monuments, fifty state forest picnic areas, twenty-eight state recreational parks, and five state-federal parks. Over 3,100 camps on state forests are leased to individuals, clubs, and other organizations; and summer homes, cabins, and hunting lodges are erected on these sites. The state forests have 4,000 miles of roads and over 5,000 miles of trails, which make them readily accessible. Some of the best fishing and hunting in the state is on the state forests.

Over 400,000 hunters and 200,000 fishermen use these forests annually. The state game refuge system consists of 199 refuges containing 687,000 acres. There are also ten large fish farms which produce about 600,000,000 young fish annually.

It is very evident that Pennsylvania has set an example in multipurpose forestry. The chief problem seems to be to build up a large state forest domain. The state hopes to acquire up to 6,000,000 acres of state forest land, or approximately 40 per cent of the total state forest area. It is working towards a sustained-yield management to supply sawmills, pulpmills, distillation plants, tanneries, and numerous other forest industries. In this way it hopes to restore rural prosperity and revive dismantled forest towns, both of which departed with the virgin forests. Pennsylvania well knows that a permanent social, economic, and industrial civilization depends upon the perpetuation of her forests. The state-owned forests will also help to control stream flow and provide outdoor recreation and public hunting and fishing grounds.

In addition to the five states just mentioned, seven others took up forestry work prior to 1900. These were Kansas, Maine, New Hampshire, Wisconsin, Minnesota, Utah, and Oregon. In the first decade of the twentieth century, under the impetus of federal development, there were added Connecticut, Michigan, Massachusetts, North Carolina, Washington, Maryland, New Jersey, Rhode Island, Montana, and Vermont. Several other states made abortive attempts but subsequent legislation was necessary to really get the work started.

Federal Coöperation with the States

It will be recalled that Section 2 of the Weeks Law of 1911 authorized the appropriation of $200,000 a year to enable the Secretary of Agriculture to coöperate with the states in protecting from fire forested watersheds, either state or private, on navigable streams, on condition that the state provided by law a system of fire protection and that the state contribute not less than half the total cost. At this time sixteen states had appropriated money to protect their forests from fire. Upon the passage of the act, eleven states immediately entered into coöperative agreement with the federal government, and by 1925 the total had increased to twenty-nine. In 1911 a total of about $250,000 was expended by federal, state, and private agencies on about 60,000,000 acres. In 1925 before the Clarke-McNary Law went into effect the total was about $2,500,000 expended on about 180,000,000 acres.

Between 1912 and 1923 Kentucky, Virginia, Texas, Louisiana, Illinois, Indiana, Nebraska, Tennessee, West Virgina, and Alabama organized state forestry departments largely spurred on by the possibility of securing federal funds. Such was the great impetus given both state and private fire protection by this law.

However, the Weeks Law did more than this. It established a new principle in American forest legislation which has had a profound effect upon the whole philosophy of American forestry and which is destined to lead to even greater results if carried to its logical conclusion. We no longer think of the protection of private forests as a problem distinct from that of public forests. The principle is now firmly established that forest fire protection on all lands is a matter of public concern. The burning of private forests is, of course, the concern of the private owner because it is his property that is being destroyed. It is now more than that, because it is regarded as a public concern on account of the far-reaching effect of forests upon streamflow, erosion, local industries, and social and economic welfare. If this is so, then the public must pay part of the expense of protection. But it goes without saying that if the federal and state governments meet part of this expense, these same agencies must also have something to say about the organization and the technical methods by which it is accomplished. The net result has been the raising of the efficiency of both state and private forest fire protection.

The Clarke-McNary Law of 1924 not only superseded the coöperative section of the Weeks Law but also added several new features, as we have already pointed out in the discussion of federal forestry. The first five sections of this law removed the limitation that the forests to be coöperatively protected had to be on watersheds of navigable streams, provided for coöperation through the states with private forest owners, and added coöperative work in forest planting, farm forestry extension, forest taxation, and forest insurance. These sections were ostensibly to advance the protection of forest and water resources and to aid private owners to profitably take up forestry practice.

Under these two laws the number of states coöperating in forest fire protection has increased from eleven in 1911 to thirty-nine in 1938. Federal coöperation was important in establishing twelve or more new state forestry departments. In seventeen states the protection of private forest land was started as a state activity as a direct result of these laws. Before 1925 only eighteen states were distributing trees for forest planting; but by 1942 the number of states engaged in this work had been increased to forty-four.

As a direct result of the coöperative sections of the Clarke-McNary

Law, state forestry in the South made phenomenal progress. Between 1925 and 1931 Georgia, Oklahoma, Missouri, Mississippi, South Carolina, Florida, and Arkansas inaugurated forestry departments or commissions.

The latest additions to the coöperative laws are the Fulmer Act of 1936 and the Norris-Doxey Coöperative Farm Forestry Act of 1937. The former authorizes federal funds for the purchase of state forest lands. At the present writing no funds have been made available under this law. The law undoubtedly will prove a great boon to states unable to finance an extended program of acquisition.

Congress in 1937 passed the Coöperative Farm Forestry Act, and in December, 1939, Indiana started the first work under its provisions. This act is a recognition of the importance of the 185,000,000 acres of farm woodlands scattered over 3,500,000 farms and making up almost 20 per cent of the total farm land in the nation. It authorizes the Department of Agriculture to coöperate in the solution of farm forestry problems with the land-grant colleges, universities, and state forestry agencies. Among other provisions it offers assistance in management, securing planting material,* protection against erosion, and the harvesting, using, and marketing of farm woodlot products.

State Activities Related to Forestry

Besides the strictly forestry work carried on by State Forestry Departments and Bureaus a great deal of closely related work is done in the administration and protection of other resources and services furnished by the forest. Many of these are not in charge of the forestry officials, but are administered by separate bureaus, offices, and departments. These usually have charge of fish, game, water supply, water power, parks, recreational facilities, historic sites, emergency work, publicity, and other matters. Few states take it upon themselves to care for and control all of these resources; each state has its own local problems and handles them in its own way.

The state of New York is probably typical of how these matters are handled, and it has probably one of the most complete organizations in the country for handling them. The annual report of the Conservation Department states that the Division of Lands and Forests, which has jurisdiction of purely forestry matters, administers 2,700,000 acres of state forest preserves; acquires and reforests abandoned

* In 1942 thirty-eight states received federal allotments for distributing planting stock.

U. S. Forest Service Photo

WATERING FOREST TREE SEEDLINGS, PIKE NATIONAL FOREST, COLORADO

SELECTIVE CUTTING IN SUPERIOR NATIONAL FOREST, MINNESOTA

After a government sale of timber, young trees are left for propagation. This is a vast improvement over the "cut out and get out" system.

U. S. Forest Service Photo

U. S. Forest Service Photo

PLANTING CREW, COLUMBIA NATIONAL FOREST, WASHINGTON

U. S. Forest Service Photo

SHELTERBELT PLANTING, OKLAHOMA

The Prairie States Forestry Project was one of the largest tree-planting enterprises in history. Over 70 per cent of the trees survived. The man hanging by his feet shows the strength and height of the trees.

Photo New York State Conservation Department

HEALTHFUL SURROUNDINGS AND SOUND TRAINING

After the C. C. C. camps were disbanded in 1942, many of the enrolees went into the army, for which their outdoor training was good preparation.

FEEDING TIME AT THE CAMP

From 1933 to 1942 our idle manpower was used to restore our wasting natural wealth.

Photo New York State Conservation Department

U. S. Forest Service Photo

PLANTING DOUGLAS FIR, PIKE NATIONAL FOREST, COLORADO

Reforestation in a mountainous state like Colorado prevents disastrous floods.

PACKING YOUNG FISH OVER MOUNTAIN TRAILS

These fingerlings will stock Rainbow Lake in Idaho National Forest.

U. S. Forest Service Photo

U. S. Forest Service Photo

FOREST PRODUCTS LABORATORY, MADISON, WISCONSIN

This modern building, completed in 1932, houses a research unit of the U. S. Forest Service. The organization constitutes one of the world's leading institutions for the investigation of wood products.

"SMOKE-JUMPER," LOLO NATIONAL FOREST, MONTANA

U. S. Forest Service Photo

Photo New York State Conservation Department

BLACK MOUNTAIN OBSERVATION TOWER, ADIRONDACK MOUNTAINS
The New York Conservation Department protects about 12,000,000 acres.

PINE PLANTATION, CHUBB HILL, NEW YORK
A flourishing young growth of white and Scotch pine. Though New York has one of the best conservation departments in the country, hardly more than one-tenth of its idle and unproductive land has been reforested.

Photo New York State Conservation Department

Photo New York State Conservation Department

AIRPLANE VIEW OF TREE NURSERY AT SARATOGA, NEW YORK

This nursery has a capacity of 25,000,000 trees. By amendment to the state constitution, 1,000,000 acres of abandoned farm land is being returned to forest.

SEEDBEDS AT CLEARFIELD NURSERY, CLEARFIELD COUNTY, PENNSYLVANIA

The millions of tiny tree seedlings are shaded from the heat by lath frames. Much skill and scientific knowledge are required to raise good planting stock.

Photo Pennsylvania Department of Forests and Waters

Photo Pennsylvania Department of Forests and Waters

PINE CREEK GORGE, TIOGA COUNTY, PENNSYLVANIA

Pennsylvania has about 1,700,000 acres of state forests. This state, with its mountainous topography, is ideally suited to growing forest crops.

ROAD THROUGH STATE FOREST, TIOGA COUNTY, PENNSYLVANIA

Since many state forests are near large cities, they are increasingly useful for recreation. Recent figures show that 28,000,000 people visit the forests yearly.

Photo Pennsylvania Department of Forests and Waters

Photo Pennsylvania Department of Forests and Water

CAMPING IN COOK FOREST STATE PARK, PENNSYLVANIA
Many campsites in state forests are leased to individuals or clubs.

CAMPING ON STATE PARK IN SULLIVAN COUNTY, PENNSYLVANIA
Pennsylvania acts as host to over two million campers each year.

Photo Pennsylvania Department of Forests and Water

Photo New York Conservation Department

CAMP SITE IN NEW YORK STATE

Thirty-one camping places and 700 miles of trails are maintained by the state in the Catskill and Adirondack Mountains.

REMNANT OF ORIGINAL HEMLOCK FOREST, COOK FOREST PARK, PENNSYLVANIA

Photo Pennsylvania Department of Forests and Waters

Photo Pennsylvania Department of Forests and Waters

BATHING POOL IN CHILD'S STATE FOREST PARK, PENNSYLVANIA

PUBLIC CAMP GROUND, PISECO LAKE, NEW YORK

Photo New York Conservation Department

U. S. Forest Service Photo

VIEW FROM CLIFF IN TALLADEGA NATIONAL FOREST, ALABAMA

Photo courtesy American Forest Products Industries

TREE FARM IN ALABAMA

Private timber growers are tending more and more to manage their forests for continous production. Unfortunately only a small proportion of privately owned forests get the sort of management illustrated above.

CONSTRUCTING FIREBREAK IN SOUTHERN PINE REGION

Most forest fires in the South creep along the ground, doing great damage to young growth and injuring older trees. If fire is kept out, forests reproduce themselves more rapidly in the South than in any other part of the country.

Photo courtesy American Forest Products Industries

U. S. Forest Service Photo

HEMLOCK, RHODODENDRON, AND HARDWOODS IN JOYCE KILMER MEMORIAL FOREST, NORTH CAROLINA

U. S. Forest Service Photo

JOYCE KILMER MEMORIAL

One of the finest tracts of virgin forest in the Southern Appalachians has been set apart to commemorate the author of the poem "Trees."

A TYPICAL "MOUNTAIN COVE," NORTH CAROLINA

A fine growth of oak, chestnut, and smaller hardwoods. With a clear spring flowing from under the trees, this is the sort of place that campers love.

U. S. Forest Service Photo

farm land as authorized by the Constitutional Amendment of 1931; operates state forest nurseries; protects about 11,000,000 acres of forest lands in the state from fire; carries on intensive control work against forest insects and tree diseases; maintains extensive recreational facilities in the forest preserve; and administers twenty-four historic sites.

The Division of Fish and Game manages the wild-life resources of the state; enforces fish and game laws with a force of 163 game protectors; maintains five game farms and sixteen fish hatcheries and four rearing stations as well as six game refuges; distributes for stocking thousands of game animals and birds and hundreds of millions of fish; carries on scientific studies in the field and in laboratories; and supervises commercial fisheries.

This division of the Department is very active in improving hunting in the state. The game farms produce annually 50,000 pheasants and 4,500 adult northern bobwhite quail for distribution on non-posted land. Over 1,700 native ruffed grouse were experimentally raised in the past nine years.

The Division of Parks maintains seventy-one state parks outside the limits of municipalities and scattered throughout the state. Over 20,000,000 persons visited these parks in the 1940 season. The Division of Water Power and Control supervises the water supply sources of the state; allocates the supply of all communities; manages the state-owned water power; supervises the formation of all local river regulating and drainage improvement districts; and exercises control over certain sub-surface water sources. A separate division maintains the great mineral-water spa at Saratoga Springs. The Bureau of State Publicity carries on extensive advertising and publicity designed to bring tourists and industries to the state.

It is obviously impossible to describe in detail all the various related activities carried on by the states. One activity that is well-nigh universal is that of the artificial propagation of game fish.

The state of Maine has perhaps as great a variety of game fish as can be found in any state. These include the Atlantic salmon, the landlocked salmon, the Chinook salmon, brook trout, black bass, brown trout, rainbow trout, lake trout or togue, white perch, and pickerel. The Maine Department of Inland Fisheries and Game maintains fourteen hatcheries and twenty-three rearing stations. It is said that under natural conditions not over 10 per cent of the fish live to adult size, while under modern methods of artificial propagation about 70 per cent live to reach legal size.

Modern methods of artificial propagation as they are employed in Maine have reached a high state of perfection. Maine must raise game fish on a large scale because it has over 2,500 lakes and ponds and more than 5,000 rivers and streams. While the adult fish are in their spawning beds, they are netted by hatchery attendants and relieved of their eggs and milt. The fish are returned to the water unharmed. The eggs and milt are placed together in shallow trays indoors where fertilization takes place. The fertilized eggs are washed in pure running water of the proper temperature in the hatcheries while they develop. When the fry are of sufficient size they are placed in large outdoor rearing pools where they are fed and given the utmost care. Fresh cold water is supplied them constantly. Many millions of such fry are kept in these rearing pools until they are of legal length. They are then placed in oxygen tanks mounted on trucks and transferred to their new homes. Maine and other states now raise these by the hundreds of millions every year. New York State, for example, over the period of 1932 to 1937 has distributed between 300,000,000 and 500,000,000 fish annually, consisting of over twenty species and of seven different sizes for some of the species.

Among the outstanding developments in fish conservation in recent years must be mentioned New York State's program of the acquisition and improvement of public fishing areas. Over $300,000 has been made available by the legislature up to 1940 for this purpose, and of the 1,000 miles of trout streams considered suitable, about one-half have been purchased since 1935.

The very close relation between forests, water supply, and game fish has already been discussed. In New York, as elsewhere, changes in forest cover, erosion and silting, changes in temperature and depth of streams, and the upstream migration of certain coarse fish have materially reduced the mileage of good trout waters. The public fishing grounds were to be acquired only in streams capable of reproducing naturally brook, brown, and rainbow trout, and no stream was included with less than fifteen miles of water of the right size and temperature to support an annual stocking of at least 1,000 trout per mile.

The landowners along these streams conveyed to the state by warranty deed an easement, right, or privilege of using the bank of the stream for a width of sixty-six feet for public fishing purposes. The purchaser was also allowed to install and maintain cribbing, docking, current deflectors, and retarding devices and to protect the land from erosion by mechanical means or by planting trees, plants, or shrubs. The total cost of title examination, land, improvements and other costs has been a little less than $500 per mile. The public fishing areas are located on twenty-seven streams in the principal watersheds of the state. When about 1,000 miles of these selected trout streams have been acquired and placed under management, it is thought that trout fishing facilities should be adequate for many years to come.*

* A. S. Hopkins, "New York State's Public Fishing Program," *Journal of Forestry,* June, 1940, p. 464.

CHAPTER XVI

FORESTRY PROBLEMS IN THE SOUTH

In many respects the southern states present some of the most complex and most serious forestry problems in the country.* Their forest area is so large and their forestry potentialities are so great that these states warrant separate discussion. Aside from the mountainous areas, the topographic, geological, climatic, social, and economic conditions in these states are so similar that they can be considered as a group. The character of the forests and the methods needed for maintenance and restoration are also fairly uniform for the whole region.

Most of these states were once covered with dense forests which have now been largely cut or at least culled over. Much of the South, due to soil and topographic conditions, is absolute forest land not fit for any other crop, and one-third of the absolute forest land of the entire nation is found in the southern states. Nearly six out of every ten acres in the South should be growing timber crops. Considering commercial forest land alone, the South has 44 per cent of the nation's supply, an area of over 200,000,000 acres. While the South is primarily agricultural, vast areas now in farms are too poor to afford a decent living. Such lands should be retired from agricultural use and should be planted to forest crops.

To put these huge areas into some form of production is not only a land-use problem of the very first magnitude, but it also involves many complicated financial, economic, political, and social problems which are of far-reaching importance to the future welfare of southern agriculture and industry.

In order to get a better idea of the position of the South in the

* Most of the data for this chapter is taken from *Forest Lands of the United States,* which assumes that the following states make up the southern group: Alabama, Arkansas, Florida, Georgia, Kentucky, Louisiana, Mississippi, North Carolina, Oklahoma, South Carolina, Tennessee, Texas, Virginia, and West Virginia.

whole forestry picture, let us take a quick glance at the ownership, present stand, and annual drain in these states:

Table 6

OWNERSHIP OF COMMERCIAL FOREST AREAS IN THE SOUTH, 1938

CLASS OF OWNERSHIP	ACRES	PER CENT
Federally owned and managed	12,908,000	6
State, county, and municipal	1,696,000	1
Private: farm woodlots	83,616,000	41
Private: industrial and other	104,311,000	52
TOTAL	202,531,000	100

We are struck immediately by the fact that only 7 per cent of the commercial forest area is publicly owned. Of great importance also are the data which show that 41 per cent of the forest land is in woodlands on the farms and a total of 93 per cent is privately owned, indicating that remedial measures must be largely concentrated on lands under these classes of ownership.

The South in 1938 had two-thirds of its timber resources in the form of saw timber and one-third in the form of cordwood. The former is important for lumber manufacturing and the latter for pulp and paper products:

Table 7

TOTAL STAND OF TIMBER IN THE SOUTHERN STATES, 1938

PHYSIOGRAPHIC REGION	TOTAL	SAW TIMBER	CORDWOOD
	MILLIONS OF CUBIC FEET		
Coastal Plain	39,256	26,510	12,746
Piedmont Region	38,562	24,006	14,556
Mountain	19,991	10,637	9,354
Delta	11,697	8,762	2,935
Pine-hardwoods, West	28,508	18,210	10,298
TOTAL	138,014	88,125	49,889

This total constitutes over 26 per cent of the total stand of the entire country. Of the saw timber about one-half and of the cordwood better than one-third is of the softwood variety. (See map on p. 11, above.)

In 1938 the stand of saw timber in the South by important species was as follows:

SPECIES	MILLIONS OF BOARD FEET
Yellow pines	196,059
Oaks	65,020
Red gum	27,407
Tupelo	20,358
Cypress	11,408
Yellow poplar	9,191
Birch, beech, maples	2,810
Hemlock	2,180
Cottonwood and aspen	1,695
Other softwoods	4,985
Other hardwoods	45,457
TOTAL, all species	386,570

Just how great this stand of saw timber is can be judged from the fact that, based on 1936 national consumption of all saw timber, the South alone could supply national needs for eight years. This total also represents 22 per cent of the total timber stand of the entire country. However, the stand is further important in that it includes 77 per cent of our valuable oaks, and almost all of our remaining supplies of southern yellow pines, cypress, red gum, tupelo, and yellow poplar. Furthermore, considering that this country is still a very large importer of pulpwood, it is reassuring to know that 42 per cent of the nation's principal kinds of wood used for pulp and paper are found in these states. Of our precious hardwood supply, the South possesses 65 per cent of the remaining saw timber.

The importance of southern forests is also shown in their cut of timber and cordwood. The total timber removed from commercial forests of the United States in 1936 was 13,463 million cubic feet, of which the South furnished 6,689 million cubic feet, or 50 per cent. This region produced about two-thirds of the hardwood saw timber and 43 per cent of the softwood saw timber for the entire country.

Southern Industries Dependent on Forests

The figures just quoted indicate that southern wood-using industries as well as those of neighboring states are going to rely upon this region for large quantities of both hardwoods and softwoods. Southern states are also heavy consumers of fuelwood for home use and for curing tobacco, another of the South's leading crops. The furniture, cooperage, and mining industries are also large consumers of wood. The early fruit and vegetable products of southern farms must be shipped in boxes and crates made in the South.

The most promising forest industry in the South is the newly developed manufacture of pulpwood products. The late Dr. Charles H. Herty of Georgia demonstrated that excellent newsprint paper could be made from young southern slash pine that can be grown very rapidly on the Coastal Plain. The first newsprint mill was constructed in 1936. Since then more than a score of Kraft mills have followed. The market for newsprint is very large, and in recent years millions of dollars worth of pulpwood has come into the United States, mostly from Canada. By 1939, industries consuming pulpwood represented an investment of over $200,000,000, this including not only those plants engaged in newsprint manufacture but also those making paper and cardboard products, rayon, and plastics. As employers of labor these industries are very important. Not many years ago it was estimated that the production, transportation, and handling of pulpwood, together with the labor employed in pulp and paper mills and in mills handling their converted products, required between four and five million man-days per year.

The forests of the South are also widely used for grazing. A considerable proportion of the millions of cattle and hogs in the South graze at least part of the time over some 125,000,000 acres of forest land. Much of this area has no "fence law," and is practically grazing common. Controlled grazing is rare. The relation of this practice to forest fires in the South is discussed later in the chapter. In general, the value of woods grazing in the livestock industry is questionable. Much research is still needed to determine the best methods of growing livestock in the South.

The naval stores industry, normally yielding raw materials worth millions of dollars annually, has been and is a vital factor in the economic life of the South. In the southern Appalachian Mountains, from five to ten thousand people receive a considerable income for crude drugs from forest herbs and trees. In Louisiana alone, Spanish "moss" yields about one million dollars annually. Edible nuts from wild forest trees like the pecan provide income and food in many rural localities.

Damage by Erosion and Floods

In many parts of the South floods and erosion annually take a heavy toll, often dislocating both urban and rural economy for weeks at a time. Grazing and agricultural cropping on hillsides, together with deforestation on slopes, have been the main causes. The control of run-off in the mountainous and hilly sections is vital to agriculture, water-power, and domestic water supplies. In most areas the control of fire and intensive reforestation on areas that will not reforest naturally would appear to offer at least a part of the solution. Many soil conservation practices, such as the use of soil-protective crops and contour plowing, have already been started and must be continued and increased. The distinction between forest, forage, and farm lands must be carefully made, and each used to the best advantage.

ONE EXAMPLE OF INFLUENCE OF FOREST ON STREAMFLOW AND EROSION
(YAZOO FLOOD OF 1931)

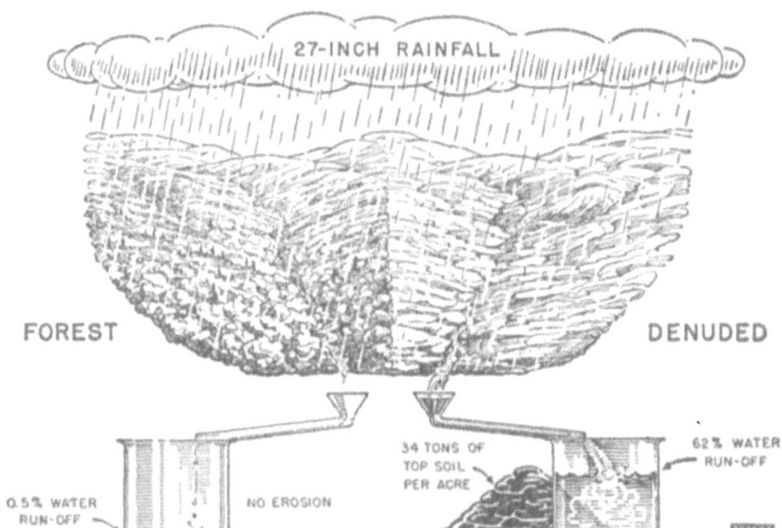

U. S. Forest Service Photo

The South is in the unfortunate position of suffering great damage from the Mississippi River, which drains the entire central part of the United States. Thus, the misuse of lands in at least a score of states has its total cumulative effect on such states as Arkansas, Kentucky, Tennessee, Louisiana, and Mississippi, which border the lower part of the river. Added to this is the destruction, almost as great, by the thirty or more smaller rivers—small only by comparison—of the South from the Potomac to the Rio Grande. The individual floods of larger rivers may take a terrific toll in lives and property damage, and though the sudden "flash floods" of minor river systems rarely get into the metropolitan newspaper headlines, their aggregate damage is very great. Thus records show that flood damage in South Carolina has averaged nearly a million dollars a year for the last twenty years, and in Tennessee the damage has been almost as great.

The Forest Service has accumulated the results of many exact experiments as to the vital rôle of forests in the control of run-off and erosion in the South. Sample plots studied in the Yazoo River basin during the 1931 flood showed that 17 inches of a 27-inch rainfall ran off sample areas of *clear* land, taking 34 tons of top soil per acre. On the adjoining *forest* land, one-tenth of one inch of the rainfall ran off, and there was no erosion. This condition prevails quite widely in this river basin, in the near-by Tallahatchee basin, and in the hilly, denuded land of northern Mississippi generally.*

The South with its developing industries is relying more and more upon hydro-electric power, because of the limited coal supply. Much of the present rural electrification and that projected for southern areas must be based upon water power; hence the control of erosion and run-off by forest restoration on watersheds is of vital importance.

The silting of water-power reservoirs has already been discussed in Chapter VI. Extreme cases have occurred in the southern Piedmont.† Thirteen major reservoirs in that region, with dams averaging thirty feet in height, have completely filled with eroded material. Their total useful life has been less than thirty years. A municipal reservoir near Spartanburg, S. C., costing $470,000, lost 17 per cent of its original storage capacity in eight years. Another near Waco, Texas, costing $2,000,000, was one-fifth filled with silt in less than six years after completion.

Such silting is a menace to water-power development all over the South. The Federal Power Commission reported that in January, 1941, the developed water power capacity of plants in the southern states was 26 per cent of the total for the entire country. Alabama, North Carolina, South Carolina, Tennessee, and Georgia ranked fourth, fifth, sixth, eighth, and eleventh, respectively, in water power production in the United States. In 1940 one-third of the electric energy produced in the United States was produced by water power, but in the nine southeastern states, due to scarcity of available fuel for steam power, between 75 and 80 per cent of the energy produced has been by water power.‡

* *Forest Lands of the United States* (House Report 323, 1941), pp. 1887-88.

† *The Land in Flood Control,* U.S. Department of Agriculture, Misc. Pub, 331, 1939.

‡ Parkins and Whitaker, *Our National Resources,* p. 330.

Forest Restoration in the Life of a Southern State

Naturally, the important question is: What share of the total employment and income in a state do forests and forest industries furnish?

Let us look at some figures from Alabama,* a state that has 60 per cent of its area in forests. In 1937 there were 62,000 full-time jobs created by forest industries, upon which about 240,000 persons were dependent for support. This was one person in eleven for the entire state. In 1929, $25,000,000 were paid in wages and salaries by its forest industries and the forest products were valued at $79,000,000.

Virginia shows even greater dependency on forest industries.† Virgin forests once covered the entire state. The forest area today is 58 per cent of the land area. Over 90 per cent of this is in private ownership, and privately owned forest land constitutes one-half the area of the state. It comprises an area nearly three times as large as that of the state's farm crop land. In 1936 about 27,500 wage earners, or 27 per cent of all wage earners in the state, were employed in handling forest resources. These workers received $19,000,000 in wages, 16 per cent of the total wages paid by all Virginia's industries. The forest products the same year were valued in excess of $109,000,000 with the value of plant equipment placed at about $49,000,000. The annual production of the nine paper mills was greater in both tonnage and value than in any other southern state.

How are the forest resources of the state being handled to maintain this important industrial position? The outlook is not encouraging. Lumber production has declined steadily from a peak of about two billion feet in 1909 to about one-quarter of this amount in 1936. There are today about one million acres of idle or unproductive land. From 1,500 to 2,000 forest fires occur annually on the protected land, which makes up only three-fourths of the counties of the state. Many more occur on the 3,500,000 acres of forest land not under organized protection. Tangible fire losses exceed $500,000 per year.

To what extent is Virginia handling its forests to insure future production? Only 3 per cent of the private forest land is handled under sustained-yield practices, while 35 per cent is being left in a partially productive condition. Over 30 per cent is being cut without regard to the future, and on 22 per cent of the forest land the condition is unknown. Cutting in general has removed the valuable

* *Forest Lands of the United States* (House Report 323, 1941), p. 638.
† *Ibid.*, pp. 55-56.

species, leaving those of low value to take possession of the ground. The flood and erosion problems of the state have steadily grown worse in the last fifty years. Wild-life resources are only a remnant of what they formerly were. The total forest resources in general are so depleted that public action is urgently needed to restore them and save many of the state's industries.

Virginia has not been selected as being particularly backward in forest restoration. At least one-half the southern states are in a position as bad or worse.

Second Growth Stands

At the present time the southern states are in possession of millions of acres of thrifty, rapidly growing second-growth stands (mostly pines), which should be forming the nucleus of a vital sustained-yield economy, if forestry is to mean anything in the near future. From present indications, it must be rescued from overcutting, especially for pulpwood, on the one hand and from devastating forest fires and the tax collector on the other. Too often these woodlots are slashed without any thought of future production.

As will be pointed out later, much of the present destructive exploitation of second growth is closely related to the meagre cash incomes of most rural families in the South. Hence any financial or economic measures which raise incomes and standards of living of the rural population in these states will materially aid in saving this second-growth resource, without which both agriculture and industry must ultimately suffer.

The commercial forest area of the South, classified by the character of the growth, as of 1938, was about as follows:

Table 8

COMMERCIAL FOREST AREA BY CHARACTER OF GROWTH *

Character of Forest	Acres	Per cent
Saw-timber areas	96,694,000	48
Cordwood areas	47,961,000	24
Fair to satisfactory restocking areas	29,114,000	14
Poor to non-restocking	28,762,000	14
TOTAL	202,531,000	100

* *Ibid.*, p. 11.

Since most of the saw-timber areas are also second growth, it is plain to be seen that most of the southern forest areas—80-85 per cent—are of this type of forest. The great importance of second-growth forests in the economic life of the South is well summarized in the Report of the Joint Committee on Forestry:

> The South is typically a region of voluntary second growth. Most sawmills are cutting second growth. The southern pulp and paper industry, annually requiring nearly 4 million cords of wood, depends largely upon it. The present-day naval stores industry is based upon second growth. Only one-eighth of the forest land is now in old growth. The United States Forest Service estimates that annual drain on the combined forest capital of saw-timber and cordwood stands exceeds effective growth by 21 per cent. Most of the second-growth stands are definitely understocked. With proper forest management and protection from fire and insects and diseases, the forest-growing stocks and the yields could be substantially increased.*

Progress in Private Forestry Since 1930 †

Of all the major forest regions in the United States, the possibilities of growing timber for profit are probably most attractive in the South. This region has most of the forest-production factors to a remarkable degree: typical forest soils, long growing season, valuable commercial species, good markets, developing wood-using industries, low-cost labor, good transportation, cheap logging, and favorable location as to export markets. Over 90 per cent of the present stand is accessible for logging. The only two business factors which the region has lacked are safety of investment (due to forest fires) and sufficient capital to invest in new wood-using industries. The first of these is slowly being achieved; the second, I understand, is being supplied in some cases by northern interests.

While the southern states are engaged in many different forestry activities, very few will be found working on all the lines of work listed. The foremost problem is fire protection, which in general has been poorly done. Most of the states administer small areas of state forests, and there are some community forests. All states coöperate with one or more of the numerous federal agencies to promote conservation work. There is some coöperative research work with lumbermen and with the federal government. Scattered efforts have been

* *Ibid.*, p. 12. † *Ibid.*, pp. 92-96.

made to reduce the tax burden. Most states give advice to private owners through the various extension services. A few states are using demonstration forests to further forestry education, and some are formulating management plans for industrial owners. There is also considerable work in the forest fish and game field and in recreational management.*

Probably the South's greatest forestry accomplishment to date has been the encouragement of private forestry. Texas reports that over two-thirds of its industrially owned forest lands are under some form of forest management. In Arkansas, ten large operating companies working on a sustained-yield basis employ foresters. The Division of Forestry of Louisiana is doing an efficient job of protecting some 6,000,000 acres of pine and hardwoods. Florida and Texas have fairly efficient protection systems. Hardwood, pine, and pulpwood trade associations are actively engaged in promoting forestry. The Southern Pulpwood Producers' Association has an encouraging conservation program.

Tree growth in the South is phenomenally fast. Some pines will grow sawlogs eighteen inches in diameter in fifty years or less. Under favorable conditions capital used to purchase cut-over lands need be invested only from ten to fifteen years before some financial returns are available. With volunteer growth already on the land even this brief period may be shortened.

Private forestry has been quick to recognize these outstanding advantages. While ten to fifteen years ago private forestry was practically non-existent, today it is going forward rapidly. The first forest plantings were an 800-acre tract of the Great Southern Lumber Company at Bogalusa, Louisiana, in 1920. But that was only an isolated attempt, for it took ten years for the total planted acreage to reach 12,500 acres. However, by 1937 this had increased to 153,000 acres. A recent Forest Service survey showed that over 17,000,000 acres of industrial forest land and about 22,000,000 acres of farm woodlands in the southern states were being handled under good forestry practices to insure the building-up of the growing stock. This amounts to 24 per cent of the total forest land in these states.

Funds for protecting private lands from fire have also increased greatly in the last decade and a half. Between 1926 and 1938 the acreage of private lands under fire protection practically doubled from 36,000,000 to 72,000,000 acres. The total appropriations for forestry by the southern states were three times as large in 1940 as in

* *The Forestry Directory* (American Tree Association, 1943), states that 14 southern states have 466 state game preserves totalling about 2,400,000 acres.

1930. The total amount of appropriations available for fire protection (county, state, private, and federal) jumped from $994,000 in 1932 to $1,666,000 in 1938, and more than 70 per cent of this was borne by the local governments. But it must be remembered that there are still 100,000,000 acres of timberland which are receiving no fire protection at all.

Now, as to the actual growing of timber. Many owners of unproductive forest land think they cannot afford to wait thirty to fifty years for a crop of saw timber, and in the South this is not necessary. In recent years, as already pointed out, there has been developing a southern paper industry, with the number of pulp and paper mills rapidly increasing. More than $100,000,000 has recently gone into the development of additional forest industries, principally pulp mills, in the southeastern section of the country. Pulpwood can be utilized down to five inches in diameter, which is much below the pole and sawlog stage. Trees nine inches in diameter can be profitably tapped for turpentine and naval stores.

A recent publication of the Forest Service * describes and illustrates some remarkable examples in timber growing on low rotations. A farmer in central Mississippi planted a worn-out hillside in 1926 to pines, and eleven years later they were twenty-five to forty feet in height and up to eight inches in diameter. Improvement cuttings had furnished both pulpwood for sale and fuel wood for the farm. In the sand hills of South Carolina a farmer planted pine seedlings in 1926, and in 1938 they were six to nine inches in diameter and thirty-five to forty feet high. A few had reached a diameter of twelve inches. On the flat, wet, cut-over "crayfish lands" of the coastal plain of South Carolina, slash pine eighteen years of age attained forty feet in height and ten inches in diameter. Many similar examples could be cited.

Large owners of cut-over lands have also achieved remarkable results. A large lumber company in eastern Louisiana reforested 30,000 acres by planting longleaf, slash, and loblolly pines between 1920 and 1930. The land was worth $1 per acre and the planting cost $4 per acre. When the trees were twelve years old the company harvested five cords of pulpwood and one cord of top wood per acre, worth $6, and there were still left on the area about eighteen cords of wood per acre. Even the cost of planting apparently can be eliminated in some of the favorable localities. It has frequently been reported that fire protection alone, for a brief period of years, will establish a good volunteer growth.

* *Southern Pines Pay,* U. S. Department of Agriculture Misc. Pub. 357, 1940.

Factors Retarding Southern Forestry

But with a few local exceptions, the whole conservation movement has made slow progress in the South, and most of this has occurred in the last fifteen to twenty years, largely with government aid. Only 7 per cent of the forest land is publicly owned. On 76 per cent of the private lands no provision is made for future timber growth. About 100,000,000 acres are still in need of fire protection. The slow progress cannot be laid to any one cause. There are many closely interrelated reasons for this retarded development. Space does not permit a full discussion of all of these, but the most important may be listed:

1. The low tax base, resulting in inadequate funds.
2. The ever-present fire menace, making forest investments unsafe.
3. The large areas in tax delinquency, making for unstable and uncertain ownership.
4. Farm tenancy with its shifting population and lack of forestry interest.
5. The plight of the small woodlot owner who is often also the border-line farmer.

Of course, there are other causes not peculiar to the South. Though southern states have pioneered the teaching of forest conservation in public schools, there has been a general lack of appreciation of the value of forest resources. Because of the commanding position which the states have held in lumber production for a great many years, the emphasis has been on exploitation and not restoration of resources. The southern states seemed not to be aware that their supremacy in lumber production would some day draw to a close.

Inadequate Forestry Funds

Funds available to the states have never been adequate for the enormous amount of work to be done in forest protection, reforestation, and forest-land acquisition. Before federal aid was available they were compelled to rely on their own limited resources. Even after federal aid came, it was not as much as had been expected. Hence today the states are still compelled to meet about 70 per cent of their forestry expenses.

The general social, economic, and financial conditions in the South seem to preclude the possibility of the southern states' solving their own forestry problems fully and adequately any time in the near future. The lower one-third of our population of ill-housed, ill-clothed, and ill-fed are mostly in the South. Recent sociological studies * of southern conditions have pointed out that this region stands lowest in the nation with regard to all the factors that make up the wealth, health, and education of a people. According to the 1930 census it has the lowest per capita wealth, lowest farm income, least farm improvements, and the lowest bank deposits and life insurance. Also it has the highest death rate, the most disease, and the least medical and hospital care. It has the smallest percentage of its population in the schools and colleges, pays the lowest teachers' salaries, spends the least on education,† and has the poorest library facilities. The per capita rural income was found to be about $200 per year, the lowest of any region in the entire country. (It is needless to point out that the South is not wholly "to blame" for its poor economic status.)

In other words, the inescapable conclusion is that the people of the South cannot raise by direct taxation large enough forestry funds and that other forms of providing funds must be worked out. It certainly would help if the federal government took over a good share of the forest lands. A harvest or severance tax on timber has already given promise of getting funds. Finally a system of federal loans to states and private owners appears to be necessary.

The Forest Fire Menace

One of the chief obstacles to forestry in the South is the old custom of setting fire to the woods for various purposes. The reasons given by the rural population for this practice are: improving the pasture, increasing berry crops, driving game, getting rid of snakes, killing the boll weevil and other harmful insects, or any of a dozen other reasons. Often no reason is assigned except the conviction that "the woods ought to be burned over every year or two."

Needless to say, such fires are destructive to young growth and to any reforestation projects that have been started, to say nothing of being a serious menace to timber, houses, barns, and cattle. The great trouble is that when once started, fires may burn uncontrolled for days, depending entirely on wind and weather as to where they will burn or what damage they will do. To engage in a large forestry enterprise with this menace abroad is simply unthinkable.

A few recent figures will show the seriousness of the situation. Of

* Howard W. Odum, *Southern Regions of the United States*, Chapel Hill, N. C., 1936.

† A number of southern states spend a larger *proportion* of their tax income on education than wealthy northern states do.

all forest fires occurring in the United States in 1936, about 83 per cent were in eleven southern states, representing 94 per cent of the total area burned in the entire country, though these eleven states have only 28 per cent of the total forest area of the country. The state foresters report that thirteen states of the South in 1941 had nearly 100,000 forest fires which burned over 10,000,000 acres of woodland and did a damage of about $20,000,000.

According to the reports of the state foresters of the various southern states the funds now being used are only from one-third to one-sixth of the amounts estimated to be the minimum required for efficient protection. Many states claim that lack of finances, not lack of public interest, is today holding back all-over fire protection. In general, outside of National Forests, all protection work is handled by state agencies, private companies contributing their share.

The significant part of this fire damage is that it is practically all man-caused and preventable. Since most southern states have good fire laws, it appears that the trouble is with the enforcement of the laws. That again is a matter of finances: to hire enough fire guards to cover the ground and arrest the guilty persons. Enforcement of fire laws also requires a stricter attitude on the part of local courts.

The forest-fire situation in the South is so peculiar to the region that it demands further explanation. In the first place, southern forests, except in swamps, are much more open in character than the dense spruce thickets of New England or the great fir forests of the West. This means that most fires in southern woods merely run along the ground, burning leaves and killing young growth but seldom killing larger trees outright. The roaring "crown fire" that destroys everything in its path, including human life, is virtually unknown. Thus the people of the region often look on the fires with indifference or approval, congratulating themselves on the death of vermin and disease-bearing germs. This harmless appearance, however, is wholly deceptive. Repeated year after year, these fires completely sap the vitality of the forest, resulting in the thin scrubby woods that are appallingly prevalent today in the South. Fertile humus is destroyed and the food supply of trees diminished. Reproduction is stopped by the killing of young seedlings. Even the older trees suffer from charred wounds in the bark, which admit insects, disease, and decay. Recent research work by the Forest Survey shows that this damage has exceeded the forester's worst fears.

The relation of cattle-grazing to southern forestry and forest fires is a complicated and fascinating story. When pioneers pushed inland from the coast they found unlimited range for their cattle and hogs in the woods. Since the only fences were those laboriously made from hand-split rails, it was far easier to fence the cattle *out* than to fence them *in*—that is, garden and corn patch were enclosed and everything outside was common ground where anybody's cattle might roam. Each owner distinguished his stock by a pattern of notches and forks cut in the animal's ear, and the leader of each herd wore a bell whose tone was easily recognized. This custom of common grazing became so firmly established that it has given way very slowly. A few states, like South Carolina, passed laws fifty years or more ago requiring cattle to be fenced—but others, like Georgia, still exercise county option, and there is still no fence law in many of the rural counties. Cattle wander over cement highways to the peril of motorists, and local law gives the cattle right-of-way. When a prominent Atlanta businessman was recently killed in collision with a cow, legislators from the cities determined to pass a state-wide fence law—but the rural districts easily voted it down. Barbed wire has replaced fence-rails—but the old hatred of fences is slow to change. John Burroughs spoke of Georgia as "this fenceless country"—and there is no denying the exhilarating sense of freedom with which one may roam the southern countryside without barrier or "keep-out" sign.

Here is where the habit of burning the woods comes in: the great majority of country people in the South are convinced that burning over the woods each spring causes grass to spring up and thus improves the grazing. As foresters have pointed out, this impression is largely a delusion—the thin blanket of green that follows a fire is mostly broomsedge, with hardly any value as forage, or coarse grasses of inferior sort. One acre of good pasture is worth 500 acres of such scanty browsing. But many of the small cattle owners have no money to put into pastures and fences, many of them have little land of their own. They wholly fail to realize that the damage done to the woods is far greater than any possible benefit to the grazing—and in any case the man who does the burning seldom owns the woods he burns. When, after years of effort, the more progressive citizens of a county manage to pass a fence law, resentment is often so fierce that those opposed to the law "set fire to the whole county" in revenge. With the best of laws, it is impossible in a thinly-settled region to watch every man who carries a match.

Heroic educational work is being done by southern foresters, so that the most remote country schools are becoming forest-fire conscious. The custom of common grazing will inevitably disappear as the region develops, and with it the last shadow of excuse for burning the woods. The whole forest-fire problem in the South is linked with the poor economic status of great numbers of people. Until this is changed, educational measures will only scratch the surface.

A recent bulletin,* emphasizing the special need of fire prevention in southern woodlands during the war, states also that uncontrolled fires affect the pocketbook of the farm woodland owner in the following ways:

1. Burning kills nature's crop of seedlings and young trees needed to establish another forest.

2. Severe fires kill some of the larger trees, weaken and slow down the growth of others and cause tree injuries by burning the bark and scorching the tops.

3. Fire damage encourages bark beetles and rot-producing diseases.

4. Fire injuries may lower the sale value of trees one-half to two-thirds. Fires often seriously damage turpentine orchards.

5. Fires frequently destroy valuable woods products already cut, such as pulpwood, logs, and lumber. Fire-scarred pulpwood is rejected by the buyers because such burned wood cannot be bleached successfully.

6. Woods fires often destroy farm buildings and fences.

Under the terms of the Forest Research Act of 1928 a nation-wide forest survey is being made, and the results are published as they become available. Three of the recent publications, covering the main forest types in widely separated southern localities, emphasize very strongly the damage done by uncontrolled fire. The Georgia study covers about 15,000,000 acres:

> Recurrent forest fires are responsible for the greater part of the loss that now occurs annually through mortality. The response of the forest to relief from periodic burning is immediate and impressive. Growth is increased, mortality decreased, and in only a few years thin stands become denser and, if seed trees are present, the openings usually become stocked with fast-growing desirable species. Protection from fire will go far toward building up and preserving the growing stock and increasing the amount of wood that annually can be turned safely into industrial uses.†

* *Preventing Destructive Fires in Southern Woodlands,* Farmers' Bull. 1926, U. S. Department of Agriculture, 1943.

† *Forest Resources of South Georgia,* U. S. Forest Service Misc. Pub. 390, 1941.

The Louisiana study covers 3,897,000 acres of mixed hardwoods:

> Despite popular belief to the contrary, forest fires substantially reduce the volume of high-quality industrial hardwood timber and the productivity of the hardwood forests... the prevention of burning will check to a considerable extent at least, the present great losses in volume and grade due to rot, insects, and windfall. It will also favorably affect the rate of volume increment and will materially reduce the loss in small trees needed for the maintenance of the growing stock*

The bulletin on southeastern Texas says this about the common ground fire:

> The shortleaf-loblolly-hardwood types are most susceptible to fire damage; reproduction and young second growth are often killed outright. Once established, longleaf pine is very resistant to fire, damage being limited mainly to scars on the stem. Fire damage to merchantable pines accounts for 10 per cent of the volume deducted for woods cull. Fires in hardwoods generally kill the reproduction and some of the young second growth. A large proportion of the woods cull of hardwoods is due to fire and the resulting decay.†

The last piece of evidence of the importance of forest fires in the South is the testimony given by F. D. Heyward, Jr., State Forester of Georgia, before the Joint Committee on Forestry in 1938:

> Probably the most striking example of the tremendous significance of forest fires in the region is shown by preliminary data, subject to revision, of the United States Forest Service survey. These data for the naval stores region compiled in May, 1937, show that the annual mortality in pine timber computed in cubic feet actually exceeded the net growth. Of this mortality conservatively 75 per cent may be attributed directly or indirectly to fires. Thus it may be stated that in the naval stores region, consisting of 44½ million acres of some of the most valuable forest land in the world, almost as much timber is burned up each year as the net amount which grows.‡

* *Forest Resources of the North Louisiana Delta,* U. S. Forest Service Misc. Pub. 309, 1938.

† *Forest Resources of Southeastern Texas,* U. S. Forest Service Misc. Pub. 326, 1939.

‡ *Forest Lands of the United States* (House Report 323, 1941), p. 95.

The question might well be asked: If uncontrolled forest fires do so much damage to the forest, why not use controlled and restricted fires to clean up dead material, improve grazing, etc.?

During the last eight or ten years the controlled burning above referred to has been given considerable study and investigation.* As a result of this work a number of foresters familiar with forest conditions in our southern pine forests have come to advocate restricted burning under certain conditions for the following specific reasons:

1. For preparing the seedbed in restocking land with longleaf pine.

2. For assistance in the control of the brown-spot needle disease of longleaf pine.

3. For discouraging competing vegetation in development of longleaf pine seedlings.

4. For reducing inflammable material and therefore the fire hazard in pine forests.

5. For improving pasture and wild-life conditions in longleaf pine forests.

These foresters give warning, however, that this kind of burning in southern pine forests requires a great deal of judgment and knowledge. Only forestry experts, thoroughly familiar with southern conditions, are capable of determining where, when, and how often the burning should be done.

A letter from the United States Forest Service to the author, dated July 9, 1943, explains very positively the position of the Service on the restricted use of fire on certain types of southern forest lands:

> The Forest Service recognizes that controlled fire may be used to advantage under certain conditions and in certain forest types in the South to obtain quicker regeneration, to reduce fire hazards, facilitate grazing, or in game management. However, the use of fire in such cases is a job for experts and should be done only under the direction of trained foresters. The Forest Service is still opposed to promiscuous burning or uncontrolled fires. It is in fact currently carrying on an intensive fire prevention campaign in the South, because fires in general do such great damage to soil, timber, wild life, and watersheds.

* Articles on the southern forest fire problem by various authors in *Journal of Forestry*, March, 1935.

The Tax-Delinquency Problem

After the timber wealth has been removed from forest lands, often the simplest thing for the owner to do with them is not to pay the taxes and let them revert to the county or the state. Many southern states have millions of acres of tax-delinquent lands on their books. How much of this tax will eventually be paid and the land redeemed and how much of the land will have to be taken over by the state for non-payment no one can predict. About ten years ago eight southern states had come into possession of some 9,000,000 acres of forest lands reverted for unpaid taxes, or 7 per cent of the total forest area of those states. This does not include some 4,000,000 acres that were delinquent but had not reverted.* Since then the situation has been growing steadily worse. Recently it was reported that in the state of Florida one-half of the forest land is tax delinquent, while in Mississippi about 1,700,000 acres recently reverted to the public agencies responsible for taxation.†

Tax-delinquency in its various stages retards forest development in a number of ways. First there is the uncertain status of a large part of the forest land. For years nobody knows who owns it. This of course seriously delays getting proper management methods started. The loss of revenue to the political units involved is also a serious matter. The necessary funds for roads, schools, and other public necessities are not collected and the rural population suffers. Finally, in order to raise the funds for public expenses the burden is transferred to the people who do pay their taxes, including those sections of the rural population that are least able to stand the burden.

Farm Tenancy

It is unfortunate for both agriculture and forestry that so large a per cent of southern farmers are tenants. The instability of this group discourages any interest in the future of the farms they occupy or in community affairs. Such farm tenancy is not conducive to fire protection, reforestation, or the handling of woodlands on a permanent basis. This kind of farm economy, little more than a hybrid between slavery and feudalism, makes a sustained-yield forestry practice impossible. It also offers little foundation for prosperous communities with good schools, churches, and industries—communities which must be the very backbone of permanent democratic living.

* *National Plan for American Forestry* (Sen. Doc. 12, 73rd Cong., 1st Sess., Washington, D. C., 1933), pp. 877-78.

† *Forest Lands of the United States* (House Report 323, 1941), p. 99.

The effect of farm tenancy on the southern woodland problem is therefore quite obvious. That the problem is a tremendous one is shown by the fact that of the total area of 139,000,000 acres in commercial farm woodlands, the South has 84,000,000 acres, or 60 per cent. Throughout the South, poor or mediocre farm lands and those abused by short-sighted agricultural practices are almost invariably occupied by tenants. The depleted condition of this farm land puts that much more drain, by either the owner or the tenant, on the farm woodlands. Incomes not forthcoming from the farm are squeezed out of the woodlands, often sacrificing the woods at any price and leaving them in ruined condition.

Farm Woodlands

Even those farmers who own and operate their own farms are often faced with conditions which make early liquidation desirable. Low prices for farm crops places the burden on the woodland, and low prices for woodland products, especially pulpwood, cause the southern farmer to question seriously the desirability of handling his woods for continuity, or to plant it to trees for future crops. Too often the price to be paid for his woodland products is dictated to him by the buyer. The help of extension foresters is obviously needed to get him the best markets for his products. The ever-present fire danger also militates against his holding areas of young growth for future harvest, especially in unprotected localities.

The farm woodland problem in all its ramifications is one of the most important in the South. Over 40 per cent of present forest ownership is in this form. The U.S. Bureau of Agricultural Economics reports that timber products cut and sold from farms in southern states in 1930 ranked in fourth place among all agricultural crops as a source of cash income, about $82,500,000. This important class of ownership must be aided by universal fire protection, help in managing and marketing through the extension services, trees at cost for reforestation, the lowering of the tax burden, and the retirement of some of the lands too poor to produce a living and the re-establishment of the farmer on better lands.

Solving the Forestry Problems of the South

Obviously, no action along one narrow front can solve the many and interacting issues facing southern forestry. The attack will have to be a sustained, coöperative attack on many fronts involving not only the local population, but also the communities, the states, and the federal government. The states cannot solve their problems without this concerted effort; their help, however, in the whole effort is indispensable.

On the matter of bringing about more equitable taxation and raising more funds, the states in some cases have already passed and enforced severance taxes on timber cut. The Mississippi Forestry Association thinks that the passage of the Timber Severance Act of 1940 in that state marks the greatest accomplishment for a conservation program that the state has ever witnessed. Commercial timber operations are subject to a 3 per cent severance tax. Arkansas has a tax of 7¢ per thousand board feet, and in 1941-2 it brought in $109,000. Louisiana has a general severance tax from 7 to 26¢ per thousand feet depending upon the kind of timber cut, and a special 6¢ tax on reforestation contract lands. Some states have a yield tax on lands under special contract, which, of course, is uncertain as to income, because the contracts are not compulsory. Alabama has an 8 per cent yield tax on land devoted to forest culture and under contract as auxiliary state forests.

As a method of raising needed revenue the severance tax in one form or other may be a part of the solution. However, it benefits principally the localities that still have large quantities of timber to cut and it does not help the depleted states that really need most of the help.

Another broad line of attack is the increased ownership of forest lands by the federal government, the various states, and their smaller political subdivisions. That forestry in the South must eventually be financed at least in part by federal funds is almost a foregone conclusion. The federal government will no doubt continue to buy outright large tracts of forest land. The other form of aid will no doubt consist of self-liquidating federal loans to states and to private timberland owners so that they may buy up large acreages of private forest lands and put them in a productive condition.

Today 52 per cent of the commercial forest area of the South is owned by large and small private and industrial holders; 41 per cent by farmers as woodlands; and 7 per cent by federal, state, and municipal authorities. The Forest Service * reports as a desirable goal the acquisition of 97,000,000 acres in the South, of which the states' share would be 20 per cent and the federal share 80 per cent. That would make the ultimate distribution of some 200,000,000 acres of commercial forest land in the South roughly as follows: 50 per cent

* *National Plan for American Forestry* (Sen. Doc. 12, 73rd Cong., 1st Sess., Washington, D. C., 1933), p. 1298.

federal, 10 per cent state, and 40 per cent private. Such a program should not be carried out, however, until private enterprise has had every opportunity to make its maximum effort. The more forest land the private agencies can handle the less will have to be acquired by public authorities.

If the suggested program were put into effect, it would mean the acquisition of about 50 per cent of the present privately-owned forest lands. In favorable areas of good species, rapid growth, suitable climate, easy logging, and good markets, more lumber companies (like the Urania and Frost companies in Louisiana) will go into continuous-yield management on a large scale. Such companies will eventually own the bulk of the private holdings. The uncertain status of millions of acres of forest land (through tax delinquency and possible later redemption) will to a great extent be gradually cleared up by wholesale purchases. Much of this land has been ruined by lumbering and fire, so that private owners would not consider it worth holding. Restoration of growing forests will fall upon the state and the federal governments which acquire them. Only the best of the private woodlands will survive as such. Most of the low-grade woodlands will therefore probably have to be taken over ultimately by public purchase.

The main lines of improvement and the legislative channels through which these may be brought about can only be briefly outlined. Most of the necessary legislation already exists, though in some cases it needs to be modified or supplemented. There is greater need for more financial backing and more aggressive action under the laws we already have:

1. Fire protection to cover all private land, involving more adequate funds under the Clarke-McNary Law. Also a change in this law to provide for distribution of planting stock to timberland owners and operators.

2. A forest credit system of low-interest revolving loans to prevent the liquidation of timber properties and to help in reforestation and in the management of forest properties for continuity. The adjustment of taxation to ease the burden of holding timber for future cutting may be handled by this method also.

3. Education through the various farm extension services to help the small owner to grow and manage his timberlands and help him find a good market for his forest products. In some cases coöperatives would be desirable, consisting of many small owners. More extension help is needed under the Norris-Doxey Act.

4. The acquisition of state forest lands could be greatly aided by grants under the Fulmer Act, since many southern states are financially unable to carry the burden. The law provides the return of 50 per cent of the purchase price to the federal treasury when the timber is harvested.

5. To help control erosion and floods, allowances should be continued under the Soil Conservation Acts. Sub-marginal farm lands should be retired under the Bankhead-Jones Farm Tenancy Act, and where necessary should be reforested and added to state forests or other units owned by the various states.

6. Enactment of a forest restoration bill whereby relief funds may be used for reforestation, the federal government to recapture at least 50 per cent of the cost when the timber is harvested. This would be valuable as a post-war measure.

7. More purchase of forest lands for National Forests in the South through added appropriations under the Weeks Law, the Clark-McNary Act, and others.

8. Large increases in research work on southern conditions, especially with regard to southern hardwoods. This would come under several existing laws.

9. Post-war restoration of the Civilian Conservation Corps, integrated in some way with our state educational systems. Public education in conservation should be greatly increased in the grade and high schools and in the teacher-training institutions.

10. Possible regulation of timber cutting through leases and co-operative agreements with private forest land owners.

The carrying out of these constructive forestry measures on 60 per cent of the total land area of the southern states will not only save the forests of that region but it will also perpetuate the forest industries. Furthermore, it will raise the standard of living of the whole people. Forest wealth, greatly augmented, will then mean continuous jobs, payrolls, and dividends. More steady income means greater sustained purchasing power. The control of erosion and floods will help to stabilize industries depending upon water. Control of grazing and of wild-life and recreation management will benefit the people in general. No other southern resource could do as much for the southern people.

Bad employment conditions in the rural South have recently been augmented by the removal of large acreages from cotton production. Such conditions emphasize the need of a new and permanent rural industry like forestry. It has been conservatively estimated that good forestry practices in the South could double or treble the production of wood products and could increase by five times the number of employed.

THE SOUTH AT THE CROSSROADS: DEPLETION VERSUS RESTORATION

Mounting requirements for lumber and other forest products during the present war have resulted in destructive cutting in the second-growth forests of the South. In some cases owners are reverting to the exploitation methods of thirty years ago, and the clock of good conservation practices is being turned backwards. A question that is very important to the South is: Are the southern states going to allow the slaughtering of valuable second-growth timber as they did the original virgin growth?

The South certainly has plenty of examples of forest depletion to look at. This effect of reckless cutting on the economic and social life of the rural population is very evident in a number of localities. Typical of the southern Appalachian country is the locality comprising twenty-six eastern counties in Kentucky. Once heavily timbered, about 75 per cent is still suited for forest land but growing only a fraction of what it should produce. When the original timber was cut out and mining disappeared, local people tried farming the cut-over land, but without success. In some counties farm tenancy doubled. During the pre-war depression, 25,000 out of 63,000 people in Pike County went on relief. The restoration of the forest and the reëstablishment of wood-using industries offers the best solution for social betterment.

In contrast to this picture let us look at a present-day sustained-yield forest operation in the South, comprising about 200,000 acres and with an annual payroll of about $250,000. Employment is provided through a program of complete utilization of the forest resources. Lumber, crossties, poles, piling, naval stores, and pulpwood are produced. During the depression neither the company nor the dependent labor suffered any hardships. The future looks bright to this community because forests are managed so as to yield a constant yearly return, and employment is held at an even keel. There will be no "boom" followed by a "bust."

Finally let us look at a public sustained-yield operation. The Osceola National Forest in Florida is one of many examples of the community benefits from public forest management. The federal government acquired 160,000 acres for about $800,000. The first problem was to rehabilitate the badly depleted growing stock. After ten years of federal management, the value of the resources doubled, though it is still producing only 50 per cent of its capacity. In 1938 it provided a payroll of $215,000 that supported nearly 2,000 people. Already, receipts very nearly balance expenses. More important still, the contributions to the local government in lieu of taxes exceeded the amount of taxes collected prior to the purchase of the land by the federal government.

CHAPTER XVII

COMMUNITY FORESTS

An extremely neglected field of forest ownership and administration is that of forests owned and operated by the smaller units of our government. In the United States, the possibilities of this class of ownership have scarcely been explored.

Community forests had reached a high state of development in the more densely populated countries of Europe before the present war, and if this European experience means anything, this type of forest is destined to play an increasingly important rôle in American forest management, especially in our more densely populated states where conditions are comparable with those in Europe. Cities, villages, or communes administered 16 per cent of the forests of Germany, 20 per cent of those of France, 50 per cent of those in Bulgaria, and 67 per cent of those in Switzerland. While the acreage of community forests in this country may never reach such a high percentage of the total, there is every reason to believe that in portions of the East and Middle West their relative importance will greatly increase.

Recent statistics seem to bear this out. The Copeland Report listed 800 community forests in 1933. By 1939 * this number had risen to 1,536, covering about 3,000,000 acres. Such forests have been established by villages, townships, school districts, towns, cities, boroughs, and counties in thirty different states, and more than 150,000,000 trees have been planted on them.

Most of the community forests are in the northeastern and Lake states, distributed as follows in certain states which rank near the top: New York 579, Massachusetts 177, Wisconsin 174, Pennsylvania 134, Michigan 112, New Hampshire 102, Vermont 44, and North Carolina 37. The New England States have 372, the Lake states 292, and ten southern states have 86.

The first states to authorize the acquisition of community forests were Massachusetts in 1882, and again in 1913, New Jersey in 1906,

* *Community Forests,* U. S. Forest Service, 1939.

Pennsylvania and Wisconsin in 1909, Minnesota, New Hampshire, and Indiana in 1913, Vermont in 1915, and New York in 1912 and 1929. In some states like Ohio, Connecticut, and North Carolina, many of the small political units are in possession of considerable forest land which has not been consolidated and placed under administration. A great many forests planted on municipal watersheds in such states are under the jurisdiction of the local water boards, and it is hoped that some provision will be made to turn these forest lands over to the proper forestry departments for purposes of organization and administration. Many such forests are now badly in need of improvement cuttings.

Benefits of Communal Forests

Experience has shown that communal forests are capable of performing many useful services for the communities that own them. In the opinion of the author these services need to be emphasized. They are so important and have such a vital bearing on current social and economic problems that every effort should be made by the federal government and the separate states to urge upon communities the acquisition and administration of such forests. As will be shown below, such a policy might help solve future problems of unemployment, relief work, seasonal lay-offs, burdensome taxation, and recreational opportunities, to say nothing of the production of useful forest products. The important uses of such forests are:

1. The protection of municipal water supplies from floods and erosion. Over 60 per cent of our population depends upon municipal water supply systems.

2. The proper utilization of lands now held or to be acquired in the future through tax reversion.

3. Provision of recreational facilities for the community, urgently needed in the more populated states.

4. The growing of timber and other products for profit, not only to ease the local tax burden, but also to provide cheap forest products for the locality.

5. Provision of jobs for the unemployed, or for those only partially employed, and of fuelwood for those on relief.

6. Provision of food, shelter, and suitable habitat conditions for wild game, fish, and birds, where these need to be protected.

7. Provision of hunting and fishing grounds where protection is no longer needed.

8. Preservation and development of areas notable for scenic beauty, historic interest, or plant and animal life.

9. Demonstration of practical forestry for the public and provision of laboratory facilities for studying biology, forestry, and other sciences.

10. As a supplement to state forests, especially in states where for financial or other reasons the creation of state forests has not been commensurate with the need for them.

How Community Forests Are Established

Four significant circumstances have already practically compelled these smaller units to administer community forests. In the first place many towns, cities, and counties already own property such as parks, "poor farms," and property connected with correctional and other institutions which have forest growth on them. Secondly, in much more than half the states tax-delinquent lands revert not to the states but to the towns and counties in which the properties are located, and it is incumbent on the local units to make the best use of them. Thirdly, hundreds of communities, especially in New England, the Middle West, and the middle Atlantic states, secure their domestic water supplies from reservoirs, lakes, and ponds, the purity and cleanliness of which depends upon the maintenance of an adequate forest cover on the various watersheds. Fourthly, in New England quite a number of the unallotted royal grants of land made to encourage early settlement actually were the first community forests on the American continent.

Thus it will be seen that long before any legislation was passed enabling the local political units to acquire forests, four different kinds of forests were automatically acquired to form the nuclei for more extensive community-owned woodlands.

By far the larger portion of the total area of town and municipal forests have been acquired for the protection of domestic water supplies. On these areas commercial use has been restricted or even entirely prohibited. The forests and plantations have been carefully protected from fire, insects, and trespass. Second in importance are those established for park and recreational purposes, many near the large industrial centers. On these areas, many of which contain lakes, swimming and boating are enjoyed in summer and skiing and skating in winter. Many types of recreational facilities must necessarily be made available for the working people in view of a universal shortening of the hours of labor in many industries and the consequent increases of leisure time. Community forests may in this respect play a rôle of far greater importance than they would if grown and managed merely for commercial use.

With a few local exceptions the town and municipal forests so far established are not now, nor do they promise to become in the near future, important sources of revenue and employment to the political units that own them. There are many reasons for this. Many of them are young plantations, not yet mature enough to cut. The cut-over mixed second growth occurring on a number of town and municipal forests will require years of careful management to bring them to their full growing capacity. About forty professional foresters are employed in community forests throughout the country. Where forests are too small municipalities will have to seek advice from the state forestry departments. In some New England states weeding and thinning has recently been done under the supervision of state officials. In short, the areas of town and municipal forests will have to be greatly enlarged and there will have to be a radical change in conception and administration to make them of any real importance for timber production.

The extent to which community forests have been started to provide wild-life sanctuaries or hunting grounds is not known. The possibilities are unlimited. Fish and game clubs in several New York counties have started forests on their own lands and have enthusiastically assisted the promotion of town and county forests in coöperation with Boy Scouts and school children. With the increase of leisure due to shorter working hours the need for accessible hunting and fishing grounds will become increasingly apparent to communities. It certainly would be to their financial advantage to encourage these sportsmen to spend money for licenses and equipment locally rather than to take the funds to other sections of the country.

By no means the least important incentive for establishing community forests should be to preserve areas of scenic or historic interest. In a number of cases attractive lakes, waterfalls, and chasms have been acquired as part of town or municipal forests for the pleasure and health of the local population. This certainly comes under the welfare function of government. Its possibilities have not yet been fully realized.

Communal Forests in Europe

European forestry practice is the result of centuries of gradual evolution and has become thoroughly systematized and firmly established. While conditions in this country differ in some respects from those in Europe, nevertheless physical, biological, and economic laws are more or less the same the world over. Hence many of the general principles of forest management and of forest administration developed in Europe are of universal application, provided the basic conditions are similar. Americans should therefore be interested to know the development and present status of communal forests in Europe, especially in Germany and Switzerland. These forests are models which we may well seek to emulate.*

A glance at the ownership of some 30,000,000 acres of forests in Germany (exclusive of Austria) shows 48 per cent privately owned, 32 per cent owned by the states, 16 per cent owned by the communes, and 4 per cent owned by institutions and associations. No forests are owned by the national government. This forest area comprises 27 per cent of the total area of the country. The states, besides controlling the management of their own forests, also take an important part in the management of all communal and some of the private forests, so that a total of about 70 per cent of the forest area is actually under state control.

The advantages of state ownership and control are very evident when one compares the greater annual yield of state forests and state-managed communal forests with those under private control. The state forests yielded in pre-war days an average of 58.5 cubic feet of wood per acre, the communal forests 53.1 cubic feet, and the private forests only 40 cubic feet. It is interesting to compare these figures with the estimated average yield of American forests, which has been computed to be about 14 cubic feet per acre. Of equal importance to us is the experience in Germany that the more strict the supervision by the state, the greater is the return per acre. Likewise, the greater the expenditure in the practice of forestry (for the necessary improvements) the greater is the net return per acre. As usual it takes money to make money, and this is no place for parsimony.

The communal forests of Germany are a natural result of the development of property conditions in that country. Gemany was originally settled by tribal chieftains, and their land was not divided but was owned as a whole by the tribe. It was not until the sixth century that one finds the beginnings of private property in woodlands. Following the Thirty Years' War a realignment of property rights materially reduced the communal holdings. But down through the centuries a large part still remained communal, for the aggregate of communal forests is still between five and six million acres.

The land owned by each commune (mark) was managed by and for the entire community. Each "marker" or share holder was permitted to cut as much timber as he needed for his own use. But along with these timber rights there sprang up, in time, rights of the chase, of

* For the purposes of this historical sketch, the change and devastation wrought

pasturage, of gathering forest litter and firewood, and of collecting other forest products. These eventually proved very difficult for cities and villages to eradicate, and even today many state and municipal forests are encumbered with these "servitudes." The history of German forest policy in this connection certainly sounds a note of warning, lest a similar practice in our own country lead to a similar situation later. At present we are granting "free use," pasturing, berrying, and other privileges which will be just as difficult to eradicate when more intensive forest management is inaugurated.

This communal ownership of forest land, in which each shareholder would help himself, worked successfully as long as plenty of timber was available. Gradually, the necessity of restrictions became apparent and officials were appointed to supervise the felling, removing, and use of the timber. Provisions were early made for planting trees to replace those cut. As a result of these and other restrictions, a practical system of forest management developed in these communal forests by the end of the fourteenth century—that is, about 100 years before Columbus set sail for America.

Germany has often been held up as a model of what can be accomplished by scientific forest management. There is hardly a policy, a method, or a practice that has not been tried there. Most certainly communal forests have been tried out. Germany possesses some of the best examples of municipal and communal forests in the world, many dating back almost 600 years.

As early as the fourteenth century, timber shortages threatened in certain localities and prompted the first attempts at planting communal forests. In 1368 the city of Nuremberg started a forest by sowing several hundred acres to pine, spruce, and fir. Incidentally, this forest was started by one Peter von Strohmer, a distant relative of the author of this book. The municipal forest of Heidelberg has been under administration continuously since 1392. This forest now comprises over 7,000 acres. The municipal and communal forests of the famous "Black Forest" district of Germany are known the world over. Berlin has over 58,000 acres of forests within the city limits; Frankfort-on-the-Main has over 10,000 acres inside the city; and Görlitz has the largest municipal forest in Germany, comprising over 72,000 acres.

These forests are important sources of revenue to many German cities, not only reducing local taxation but even in many cases making such taxation entirely unnecessary. Timber production is among the main objectives of these forests, and intensive methods of management yield high returns. The average net profit is given as about $5 per acre per year. It is said that several hundred German towns have forests that yield sufficient revenue to pay all local public expenses and in some cases an annual dividend to their citizens. Communal forests are also used for hunting, fishing, and other recreation. Sometimes reserves of timber are set aside to finance unusual public expenditures, like the building of new schools.

Communal forestry in Switzerland is of particular interest to Americans because such a large proportion of the forests of this country belong to the communes, because of the exceptionally high net returns per acre, and because of the development of an efficient system for the protection of mountain forests.

Switzerland has about 2,400,000 acres in forests, which comprise 21 per cent of the area of the country. The communes own 67 per cent of these forests, private individuals 29 per cent, and the cantons or states 4 per cent. The federal government does not share in the ownership of forests, but it does have a great deal of supervision over their management.

America can well profit by the experience of the Swiss, because the bulk of their forests are protection forests very similar to those found in the mountainous districts of our own country. The Swiss have succeeded in proving beyond a doubt that highly profitable forest management can be maintained side by side with the protective function. These forests furnish protection from erosion, floods, snowslides, and landslides in a country where the topography for the most part is exceedingly precipitous. All such forests, whether in public or private ownership, are under strict federal supervision as to cuttings, and the clearing of forests is prohibited except by federal permission. All communal forests are under the general supervision of the cantonal (state) governments.

The average annual growth on Swiss forests in pre-war years was over 50 cubic feet per acre and was highest in state forests. Due to over-cutting during the First World War, the annual growth was reduced, but now in general the annual cut is less than the annual growth in order to restore the balance. Financial returns from communal forests average between $3 and $9 per acre per year; but during one year the net receipts of the town forest of Winterthur reached $11.69 per acre. One of the most noted of communal forests is that owned by the city of Zurich, the records of which go back over 1,000 years. It contains 4,200 acres and yields a yearly net revenue of between $20,000 and $30,000. It is a wonderful example of orderly, systematic, scientific forest management.

It must always be borne in mind that the above yields are from land too steep and often too rocky to plow. It is the lowest grade of land which these countries possess and probably would be of no use for any other purpose than forestry. But in spite of this fact the yields compare favorably with those secured on many of our cultivated farm lands.

Communal forests developed in France very much as they did in Germany. They still are the most important form of public forest ownership in France. Of particular interest are the so-called taxless villages of France. In the Jura Mountains much of the spruce and fir timberland is owned by communes who are in the wood-cutting and wood-selling trade. These forests are managed in a businesslike way so that quite often there is considerable profit after all expenses have been paid. The official records show that twenty-four French communes in the Jura Mountains, with a total population of over 10,000, not only pay no local taxes but actually receive annual dividends of from 100 to 200 francs. There are also eighteen other communes with a population of 8,000 where no dividends are paid but where no local taxes are necessary. In most of these communes each family also receives free firewood, and in some cases free sand and stone for building.

Development in the United States

It is perhaps very fitting that the original impetus to the community forest movement in this country should come from New England towns. These miniature commonwealths, for many years during the early colonial period, were the only political corporate bodies in the region, having preceded both the state and the county governments. Even after these larger political units appeared, the town with its efficient, truly democratic system of self-government held its ground. It exercised a practical sovereignty over property and persons based upon thoroughly democratic principles. In the town meeting all local public projects and activities are still openly discussed. How the taxes are to be raised and how the funds are to be spent are matters determined by the vote of the people. With the actual operation of government in the hands of the taxpayers it is easy to see why they would vote for a project like a town forest which can be made such an intimate part of local household economy.

New England Town Forests

The earliest town forests are parts of the unallotted royal grants of land given to encourage settlement in early Colonial times. While the agricultural lands were divided among the settlers, some of the forest land remained common property. Thus without any specific state legislation a number of town forests were in existence in the early part of the eighteenth century. The town of Newington, New Hampshire, boasts of a town forest of 122 acres which dates back to 1710. Danville, in the same state, has managed a town forest of 75 acres since the time the town received its original charter in 1790.

Most of New England's town forests, however, date from 1913 or later, for beginning at this time laws were enacted which specifically encouraged this type of ownership. The Massachusetts Town Forest Act of 1913 was the first of these, and the city of Fitchburg, Massachusetts, claims the distinction of having established the first town forest under state law in the United States. Up to 1939 some 372 town and city forests aggregating nearly 200,000 acres have been established in the New England states.

Unquestionably the greatest development of the New England town-forest idea has been in Massachusetts.* This is a wealthy and populous state with good markets on every hand for a great variety of wood products, and with abundant need for recreational and watershed forests. In addition it possesses an active State Forestry Department and a highly enthusiastic and efficient Massachusetts Forest and Park Association. This combination, together with that proverbial New England intelligence and thrift, spells success. In view of this set-up of motivating forces we shall look into the Massachusetts situation in detail in the hope that what it has done in establishing town forests, many other states can most certainly do where conditions are similar.

The Town Forest Act of 1913 permits towns to acquire by purchase or gift lands for the purpose of forestation, and the State Forestry Department may furnish such towns free of charge with tree seedlings for planting such lands. A town, by a two-thirds vote at an annual town meeting (or a city by a two-thirds vote of the city council), may determine to take or purchase lands within its limits and may appropriate money and accept gifts of money and land to acquire a public domain to be devoted to the culture of forest trees, or to the preservation of the water supply. In connection with acquiring such lands the town or city may exercise the right of eminent domain.

The management of town forests is in charge of a special Town Forest Committee appointed by the selectmen and composed of three members who serve for staggered terms of three years each, without

* See *The Town Forest in Massachusetts,* Bull. 159, Mass. Forest and Park Association, 1937.

compensation. Other acts pertaining to town forests have to do with erecting and leasing buildings on such lands; limiting the expenditures for land and buildings to not more than the appropriation allotted; and limiting the town indebtedness for this purpose to not over one-half of one per cent of the last assessed valuation of the town or city, payable within a period of thirty years.

As already pointed out, the city of Fitchburg established the first town forest of 109 acres under this act. But the movement for town forests actually did not get under way until 1923, and in the last fifteen years the growth of the movement has been little short of phenomenal. At the beginning of 1923 there were only four town and city forests entered under the act; by the close of 1939 there were 177, aggregating over 170,000 acres, on which over 5,000,000 trees have been planted. This includes 75 watershed forests with over 40,000 acres of forest land or potential forest land owned by cities and towns outside of metropolitan Boston for the protection of their water supplies and which have not yet been placed under the act. Besides this area, the cities and towns of Massachusetts have within their limits idle forest land suitable for town forest purposes aggregating about 500,000 acres.

For the splendid results so far accomplished, credit must go largely to the Massachusetts Forestry Association (now called the Massachusetts Forest and Park Association). As soon as the Town Forest Act was passed, this Association requested each town to secure the appointment of a special committee to study the possibilities of establishing town forests. Through their efforts many communities and towns were persuaded to establish such forests. Almost every means for education and propaganda was used to advertise their advantages. Most ingenious of these was the offer to plant 5,000 trees free of charge for any city or town which would legally establish a town forest of 100 acres or more. This brought satisfactory results, for the Association down to 1936 has planted 230,000 trees for forty-six municipalities. Private organizations in various parts of the state made similar offers. County forestry associations, city chambers of commerce, county improvement leagues, and box companies either offered prizes or offered to plant 5,000 trees for any towns establishing forests within their respective counties.

The Forest and Park Association in 1929 organized the first conference of town forest committees, and conferences are held annually. This organization encourages the exchange of ideas among the various committeemen, and the meetings take the form of field days on the town forests that have shown the greatest progress.

The Division of Forestry of the state undertakes to prepare a working plan for any town forest, and a number of town committees (notably Fall River and Westfield) have taken advantage of this offer to follow a definite plan in the management of their forests.

It might be of considerable help to communities desiring to create town forests to know how the above-mentioned special town committees went about the task of securing forest land. When these committees set about finding suitable forest lands, they first looked over lands already owned by the towns. These lands usually consisted of those used for the protection of the water supply, the town farms, tracts that had been donated, areas that had reverted to the town for unpaid taxes, or lands acquired for a specific purpose but no longer needed. If no such lands were available, owners of large areas of woodland or of cut-over land were asked to make gifts. Failing in this, the committee resorted to outright purchase. Thus one-fourth of Massachusetts town forests were established on old town farms, one-fourth on watersheds protecting town water supplies, one-fifth by gift, one-fifth by purchase, and the remainder on tax land or on areas originally acquired for other purposes. The majority of the town forests comprise from 100 to 500 acres. Some of the largest are those of the towns of North Adams (6,720 acres), Westfield (5,404 acres), Russell (3,000 acres), and Fitchburg (2,067 acres), all of which are essentially watershed forests.

The three outstanding purposes for which Massachusetts town forests have been created are: watershed protection for water supplies of municipalities, the utilization of idle forest land, and for recreation. But the forests have yielded many other by-products and services. To some considerable extent they have been a source of employment, and the vast amount of thinning, pruning, weeding, and other operations needed indicates that the town forests will continue to take care of some of the unemployed. The welfare departments of many cities and towns have secured firewood for the poor from these forests. Many Massachusetts towns have established game and bird refuges to increase the supply of game in the region. Through the efforts of a local garden club, a wild flower sanctuary has been established by one town. Towns and cities have secured shade trees for street planting and trees and shrubs for ornamental planting. Even high school athletic fields have been located on these areas. Thus it will be seen that town forests are an established institution in Massachusetts and promise to become an integral part of the state's policy to make better use of its forest resources.

A recent report of the State Planning and Development Commission of New Hampshire, dealing with the state's resources, places forest wealth as her greatest asset. The predominant soil of sandy and gravelly loam is better fitted for growing timber than for any other purpose. Her forests are of paramount importance to protect recreational and wild-life values, for watershed protection, and to supply New Hampshire farms and wood-using industries with the needed wood. The report recommends that more land be devoted to scientific reforestation. It is therefore evident that town forests must play a greater rôle in her future forest policy.

The history and development of town forests in New Hampshire is much the same as that in Massachusetts, only on a smaller scale. With over 80 per cent of its area still in forests, this state, perhaps, has not felt the need for a vigorous town forest policy. Up to 1939 there were over 100 town, city, and village forests, totalling over 25,000 acres, on which over 2,000,000 trees had been planted. The purpose of many of these forests is cordwood and timber production, though the largest of them are used for watershed protection. Of the latter class there are thirteen towns and cities that own 9,500 acres upon which most of the planting has been done.

Town forests of New Hampshire have been acquired mostly by gift (thirty-three) and by purchase (thirty-three), the remainder by tax title or from unallotted lands which were the property of the town since early Colonial times. About one-half of the town forests were created before the enactment of the Town Forest Act of 1913. As already stated, the oldest town forest in this or any other state dates back to 1710.

Many town forests in New Hampshire are intimately connected with early Colonial town history and have been managed as woodlots for 150 years or more. They have for generations provided the town schools, libraries, and churches with timbers and firewood. The poor people of the town have frequently been supplied with wood, and the sale of timber has often helped to swell the town treasury. These forests were thus a going institution long before the State Forestry Department was organized in 1910. In at least one case—Danville—the town forest has helped to support the church from early Colonial times.

A number of New Hampshire towns have made considerable progress in developing their forest properties. In many cases the history of these forests is highly interesting and dramatic. The City of Manchester has acquired for watershed purposes about 3,800 acres of forest land, some of it containing stands of timber. For over twenty-five years reforestation has been going on, and to date well over

1,200,000 trees have been planted. During one recent ten-year period, over $13,000 was received from the sale of wood and timber. The city of Concord owns about 500 acres, mostly for watershed protection, and has planted over 150,000 trees beginning in 1895. The forest property is now worth upwards of $50,000, and reforestation is keeping pace with cutting. The city of Keene has acquired largely through purchase some 2,000 acres and has planted about 100,000 trees since 1920 to protect watersheds. The proceeds of the sales of wood and timber are used for purchasing new tracts, for making needed improvements, for developing a recreation forest for the use of children, and for the support of the city hospitals, churches, and charitable institutions.

Public-spirited citizens often get behind the town forest movement in New Hampshire. The town of Sunapee boasts of one of the best-timbered forests located near Lake Sunapee, comprising 100 acres of old-growth pine, hemlock, and beech. Through the efforts of its citizens contributions were solicited to save the forest for the town. The owner agreed to sell, and Dewey's Woods was finally acquired by the town. Although its value does not exceed $5,000, it possesses recreational values far in excess of this amount. The town of Merrimac snatched 100 acres of old-growth pine from the woodsman's axe in less time than it takes to tell it. Ten interested citizens advanced the money for purchase because town-meeting day was months off. Cutting was stopped after four acres of primeval forest had crashed to the ground. Today Twin Bridges Town Forest has been saved for the people, and the four acres that had been cut over were soon replanted, the state furnishing the trees and the school children doing the planting.

The Town Forest Act of 1913 authorized any town or city in New Hampshire to vote at any legal meeting, sums of money to purchase, manage, and improve lands for the purpose of growing wood and timber. These lands were to be managed under the direction of the State Forester through the town forest selectmen and the net proceeds of any sale of wood and timber was to go to the town treasury. The first conference of town-forest selectmen and committeemen in the vicinity of Concord was held at Warner in November, 1930, at which time eight towns sent representatives.

The Vermont Act of 1915 authorizing the acquisition of town forests is very similar to that just described for New Hampshire. Under this act Vermont has up to 1939 acquired forty-four town, village, and city forests totalling around 10,000 acres. About one-third of these are watershed forests and the rest have been established for growing timber and cordwood. On the denuded portions of these lands over 2,500,000 trees have been planted. About one-half the forests were acquired by purchase, the rest were gifts or former poor farms no longer needed. The largest forests are those owned by Rutland (4,000 acres), Montpelier (806 acres), Essex Junction (752 acres), Northfield (500 acres), and Bellows Falls (400 acres). All of these were established for watershed purposes.

In recent years these forests have been an important source of work for the unemployed. Thinning, pruning, eradication of currant and gooseberry bushes (to control the white pine blister rust), and the cutting of fire lines has been undertaken in the Rutland, Bellows Falls, Springfield, Essex Junction, Stowe, and other forests. The City of Rutland during the winter of 1931-2 spent between $6,000 and $7,000 on this kind of work, employing about forty men daily.

Vermont, like New Hampshire, is a state in which forests are the principal resource. About 60 per cent of Vermont, or about 3,500,000 acres, is productive timberland, on soil too steep and too rocky to plow. The future welfare of both states demands that forest land be turned to productive use. Both states already have both national and state forests, in addition to the town forests mentioned. It would seem that both states offer unlimited opportunities for the acquisition of all three types of publicly owned forests.

The other two New England states that have some town forests are Connecticut (30,000 acres) and Maine (1,300 acres), in most cases used for watershed protection in connection with city water supplies.

Community Forests in New York

The wealthy and populous state of New York has all the prerequisites for profitable forest management: cheap and abundant labor, diversified markets, abundant capital, good transportation, accessibility, and favorable climate and soil. It has made the greatest progress in acquiring state and community forests. Outside of New England, New York has the greatest number and the greatest area of community forests.* Of a total of over 1,536 such forests in the entire country in 1939, New York has more than a third. These 579 forests, with a total area of over 200,000 acres, are owned by counties, townships, villages, cities, and school districts. Fully 75 per cent are forests surrounding municipal water supplies, on which over 70,000,000 trees have been planted since 1910.

* H. F. Prescott, *Municipal or Community Forests*, N. Y. Conservation Department, 1939.

Although New York leads all the states in its area of state forests and parks under administration there is also ample opportunity to develop an extensive system of community forests, using the present ones as a nucleus. The state has over 4,000,000 acres of land, fit only for growing forests, that are denuded and idle and in the present condition an economic liability. In addition there are perhaps another 4,000,000 acres of inferior growth, so poorly stocked that it is incapable of producing a wood crop of any value and of so little present value that it scarcely produces enough income to pay the taxes on it. The sad part of this is that every acre of this could be made to yield a profit.

Let us look at the New York situation from another angle. As already stated, about one-half of the land area of the state is better suited for growing timber crops than for any other purpose. The virgin forests have long been exhausted and the second growth stands are usually of inferior quality. These in 1919 furnished less than 15 per cent of the materials used by local wood-using industries. The average cost to manufacturers in the same year was over $60 per thousand board feet, and much of this wood could have been more cheaply grown right in New York. Due to lack of accessible raw materials and to high freight rates on lumber and wood brought into the state, local wood-using industries, once the greatest of any state in the Union, have shown a steady decline in the last twenty-five years. The result has been a progressive closing down of small wood-using industries in the smaller cities and villages, with the inevitable concomitant of abandoned farms and a migration of the rural population towards the big cities and to other states. The question might well be asked, can forestry stop this exodus, rehabilitate the farms and villages, and rebuild and revitalize the rural population that has been rapidly diminishing?

Most of New York's community forests date from the enabling act passed in 1912, though some plantations date back to 1909. Up to the end of 1939 New York had 50 city forests, 215 village forests, 51 township forests, and 215 school district forests.

Most of the large cities of New York own municipal forests in connection with their water-supply reservoirs. On these watersheds millions of young trees have been planted, largely to protect them from the erosion menace. Much of this vast planting program started in the period of 1909 to 1916 and has been kept up ever since. New York City has planted over 5,500,000 trees on the Ashokan, Gilboa, and Croton watersheds. Glen Falls has planted 2,500,000, Rochester over 2,000,000, and Little Falls over 2,500,000. Troy, Oneonta, Gloversville, Middletown, Johnstown, Salamanca, and Utica have been large planters on their municipal forests.

Many incorporated villages in New York have started village forests. The largest is owned by Carthage, which started in 1910 to plant trees and has up to 1931 set out over 850,000 of them. Malone ranks second with over 500,000, and Canajoharie, which started in 1915, has planted 270,000.

About 215 school districts up to 1939 had started school forests in the rural districts where land is cheap and many of these at maturity will pay either wholly or in a large part the cost of maintaining the school. By planting some fixed number of acres every year for a series of years, a steady revenue may be derived. Such planting could very well fit in with Arbor Day exercises and might well give this day a new significance. A number of school districts already have a good start on such a program.

The largest school forest in New York was started in 1921 by the Town of Watson, in Lewis County. By 1931 over 86,000 trees had been planted by the school children on a tract of 98 acres which was purchased for $300 by a vote of the people of the school district. The planting was done in observance of Arbor Day. The trees were furnished free by the State Conservation Department.

Now, as to the enabling law which makes these forests possible. It might be well for those persons and organizations interested in conservation to call this law to the attention of your local mayor, Board of Supervisors, town board or school board. The chances are they do not know that such a law exists.

The establishment of community forests is provided by Section 60 of the Conservation Law of 1912. The law provides that a county, city, town, or school district may acquire by purchase or gift, or take over lands in its possession within its boundaries, and use the same for forestry purposes. The governing board of a county, city, town, or school district may appropriate money or issue bonds, either for the purchase of lands, to establish plantations, or for the care and management of forests. Such boards may undertake such work at regular or special meetings by a majority vote of the board after two weeks' public notice. The Conservation Department may give advice and assistance in reforesting work and may furnish the trees, provided they will be planted in accordance with instructions. The local board shall maintain, manage, and operate such forests for the benefit of the district and shall have control of the funds derived from such lands.

Experience in New York State has shown that such forests are usually started upon the initiative of one of the many progressive organizations or service clubs found in practically every community. Rotary, Kiwanis, Lion's Clubs, Chambers of Commerce, Sportsman's Clubs, Boy Scouts, American Legion Posts, Conservation Clubs, Granges, Women's clubs, D.A.R., and Parent-Teachers' organizations have been powerful agencies in sponsoring and securing the adoption of plans by the above-named governing boards. Boy Scouts and school children have given their services freely in planting the trees.

Community Forests in Other States

A number of other states have also started community forests. The cities and villages of New Jersey have about 60,000 acres. Newark has a forest park comprising 36,000 acres in charge of a forester on which 2,000,000 trees have been planted. North Carolina has about forty forests with a total of about 44,000 acres. Most of these have been acquired for the protection of municipal water supplies. On most of them no commercial cutting has been allowed and strict control over trespass has been exercised. The largest forest of this kind in the state is in Asheville and contains 22,000 acres. Pennsylvania has 134 town, borough, county and city forests, totalling about 50,000 acres, on which 5,000,000 trees have been planted. These are used for watershed protection and park purposes. No clear line of distinction seems to have been made between these two uses. Maryland has nine such forests with a total area of 24,000 acres. Baltimore has the largest—over 15,000 acres—which is under the jurisdiction of the Water Department. Ohio has twenty-five forests totalling 22,000 acres. Here again no clear line of distinction seems to be made between metropolitan parks and purely forestry enterprises.

Other community forests are scattered through the states of Alabama, Colorado, Connecticut, Florida, Georgia, Indiana, Iowa, Illinois, Kentucky, Louisiana, Maine, Michigan, Minnesota, Missouri, Rhode Island, South Carolina, Texas, Utah, Virginia, Washington, and Wisconsin. One of the largest municipal forests on the Pacific Coast is that owned by the city of Seattle, Washington, covering 67,000 acres, from which the city has derived considerable income.

County Forests

The idea of county forests is probably still in its formative stages of development. In some states it is going forward with considerable success, while in others it presents a problem evidently too great for the counties in the state to handle. According to the Copeland report of 1933, the total area of county forests was somewhat over

500,000 acres. By 1940 this total had reached about 2,000,000 acres. New York, Wisconsin, New Jersey, Illinois, and California have made some progress in county forestry.

Although New York does not show the greatest acreage in county forests, its legal setup seems to give more promise of success. In 1939 it had fifty county forests with a total area of about 25,000 acres. The Community Forest Law of Section 60 of the Conservation Law of 1912, already cited in connection with other types of community forests, also applies to counties. Up to 1929 over 18,000,000 trees had been planted on lands acquired as county forests. Some counties have adopted a policy of planting some trees every year.

In 1929 the New York legislature passed a bill giving state aid for county forests. It provides for contributions up to $5,000 a year to any county for the purchase of lands, for their reforestation, and for the protection of the plantations or other forest growth. Title of lands is vested in the county, and the area must be forever devoted to the purposes of watershed protection, the production of timber or other forest products, and for recreation. The lands are exempt from state and county taxes. The board of supervisors may sell the timber and other forest products under rules and regulations prescribed by the Conservation Commissioner. Twenty-one counties bought land with state aid, and over 20,000,000 trees were planted on this land. under the law, but in 1933-34 no state appropriations were made and the work was temporarily interrupted. Boards of Supervisors felt, on account of the economic depression, that it was wise to curtail expenditures for such work.

The situation in Wisconsin is very different. In about one-half of our states, tax-delinquent lands revert to the county rather than to the state or to the town. Wisconsin is one of those states and it has acquired county forests totalling 1,700,000 acres up to 1939, on which over 35,000,000 trees have been planted. Because of the enormous amount of tax-delinquent cut-over land in the state, the future area of county forests may reach 7,000,000 acres, according to a recent survey.

It is evident that without some kind of outside aid the financial burden of protecting and developing such vast acreages of cut-over lands by the counties would be practically impossible. Under these conditions tax lists are depleted and very little cash is taken in for taxes. Wisconsin real estate taxes are paid to the townships, which turn over the county's share to the county treasurer. He returns some of the money to the townships for school purposes. However, the townships may turn over delinquent tax lists to the county in lieu of cash. Hence, if the township is heavily delinquent, the county receives very little if any cash, but it must still pay county aid to schools in cash. In some instances banks have refused to lend the counties any more money. All of which is a very good example of what happens under the old system of "cut out and get out." When taxable wealth vanishes on a large scale, as it did in these vast forest areas, the resulting disturbances are not only economic, but financial, political, and social as well. Under continuous-yield management this would not be possible.

To remedy this situation the Wisconsin legislature passed the Forest Crop Law of 1927, which was amended in 1929 and 1931.* After the Conservation Commission lists the county-owned land as forest crop land, the state pays annually 10¢ per acre to the county and 10¢ per acre to the township in which the land lies. The county is to use its share towards the development of the forest units. The state in return is to receive 75 per cent of the yield when the forest is cut. In effect this amounts to a state loan. There is only the question as to whether the financial condition of the state will allow this to go on for from 60 to 100 or more years until the crop matures. If not, then the burden of protecting, improving, and reforesting these cut-over lands will have to be carried by federal aid.

In California, several counties (Los Angeles, Ventura, and Santa Barbara) have active forestry organizations. Some of the recent annual appropriations for forestry activities have varied from $500,000 to as high as $1,000,000. These enormous sums were expended mainly for forest fire control, watershed protection, park development, and roadside landscaping. The vital importance of the forest and brush cover in these counties for the protection of watersheds against erosion and floods gives ample reason for these large expenditures.

* Copeland *Report*, p. 847.

CHAPTER XVIII

ECONOMIC FACTORS IN PRIVATE FORESTRY

THE PLACE OF FORESTS in the household economy of the nation has already been discussed in previous chapters. While federal and state forests will always be important, at present the management of private forests is of far greater concern. Therefore let us look at some of the financial and economic phases of the problem which have an important bearing upon private timber production in this country. Some of these economic factors are peculiar to forest production, and render forest problems conspicuously different from other national problems, and forestry business different from other kinds of business. They also explain, in some measure at least, why the science and art of forest production has not been more widely applied to private forest lands.

When the first settlers reached America early in the seventeenth century, the land area of the country could be classified about as follows: *

Original virgin forests	800	million	acres
Grasslands	600	"	"
Open woodlands and arid lands	400	"	"
Desert and other waste lands	100	"	"
TOTAL	1900	"	"

After 300 years of civilization, we find the following approximate distribution:

Original virgin forests	100	million	acres
Forest lands cut over	500	"	"
Grazing lands	330	"	"
Farm crops and pasture	820	"	"
Cities, deserts, wastelands	150	"	"
TOTAL	1900	"	"

* S. Chase, *Rich Land, Poor Land*, p. 35.

U. S. Forest Service Photo

VIRGIN CYPRESS, HAMPSTEAD COUNTY, ARKANSAS

Cypress is a light, strong wood, especially valuable for its durability under exposure to weather.

U. S. Forest Service Photo

RIVER BOTTOM HARDWOODS, CLINCH COUNTY, GEORGIA

U. S. Forest Service Photo

PINE, HEMLOCK, AND HARDWOODS ALONG LINVILLE RIVER IN THE NORTH CAROLINA MOUNTAINS

VIRGIN LONGLEAF PINE, KISTACHIE NATIONAL FOREST, LOUISIANA

This famous pine once covered the whole lower South.

U. S. Forest Service Photo

U. S. Forest Service Phot

RATE OF GROWTH

Slash pine on worn-out hillside in Mississippi. Even poor land like this will grow pines for lumber and pulpwood at a rapid rate.

U. S. Forest Service Photo

YOUNG GROWTH OF LONGLEAF PINE, AIKEN, SOUTH CAROLINA

SECOND-GROWTH STAND OF OAK AND RED GUM, LOUISIANA

This 80-year-old forest is growing in an old field that was once cultivated. The furrows of some ante-bellum cotton field may still be faintly traced on the forest floor. Such forests show the practicability of returning worn-out farm lands to the growing of wood.

U. S. Forest Service Photo

U. S. Forest Service Ph

TURPENTINE TIMBER, GEORGIA

Photo Seattle Water Department, courtesy U. S. Forest Service

FIRE LOOKOUT ON COMMUNITY FOREST NEAR SEATTLE, WASHINGTON

This largest community forest in the country was acquired to protect the city's water supply, but its mature timber has sold for over $1,000,000.

RED PINE, COMMUNITY FOREST, PETERSHAM, MASSACHUSETTS

This is on the site of the old "poor farm," and it is estimated that the village will receive $50,000 from the 99-acre tract within the next forty years.

U. S. Forest Service Photo

U. S. Forest Service Photo

WHITE PINE, TOWN FOREST, NEWINGTON, NEW HAMPSHIRE

U. S. Forest Service Photo

FUEL WOOD FROM CITY FOREST, RUTLAND, VERMONT

PRUNING AND THINNING, CITY FOREST, WESTFIELD, MASSACHUSETTS

U. S. Forest Service Photo

Photo Board of Public Works, Westfield, Massachusetts, courtesy U. S. Forest Service

CITY WOODPILE, WESTFIELD, MASSACHUSETTS

CHEMICAL PLANT SUPPLIED BY COMMUNITY FOREST, WARNER, NEW HAMPSHIRE

U. S. Forest Service Photo

U. S. Forest Service Photo

SCHOOL PLANTATION, MINFORD, OHIO

This 10-acre tract was planted by students in 1932. Picture taken in 1939.

ANOTHER VIEW OF MINFORD SCHOOL FOREST

The community is proud of this achievement by and for its public school.

U. S. Forest Service Photo

U. S. Forest Service Ph

TOWN FOREST, ESSEX JUNCTION, VERMONT

PINE FOREST, TOMHANNOCK RESERVOIR, TROY, NEW YORK
A twelve-year growth. Photo taken in 1926.

Photo New York Conservation Departm

Photo by Brown, courtesy West Coast Lumbermen's Association

FIRE ON TOANDAS PENINSULA, WASHINGTON

Started by a cigarette tossed from an automobile, this fire burned for ten days despite the efforts of 3,000 men. Because of the threat of such catastrophes, many Pacific Coast lumbermen are now members of fire protection associations.

DOUGLAS FIR, MT. HOOD NATIONAL FOREST, OREGON

Billions of feet of this fine lumber have gone into fighting ships and planes.

U. S. Forest Service Photo

U. S. Forest Service Ph[illegible]

MAGNIFICENT TRACT OF WHITE PINE UNDER PRIVATE OWNERSHIP, CLEARWATER COUNTY, IDAHO

Photo by Brown, courtesy West Coast Lumbermen's Association

FIRE PROTECTION BY PRIVATE GROWERS, WASHINGTON
West Coast timber owners maintain fire trucks for quick protection. Such tree farms as these will form the basis for a future harvest.

COOPERATIVE TREE NURSERY, NISQUALLY, WASHINGTON
An example of cooperation by private companies to meet a basic need.

Photo by Brown, courtesy West Coast Lumbermen's Association

Photo by Brown, courtesy West Coast Lumbermen's Associati

BABY TREES BY THE MILLION

Work on the large nurseries is highly mechanized.

These figures show startling changes in land use. It will at once be seen that almost 90 per cent of our original virgin forests have been cut or culled over. The original grasslands have found use partly as range land for livestock and partly for farm crops and pasture. Open woodlands and arid non-agricultural lands have also been used partly for grazing. More than one-half of our forest lands are also used for grazing. In fact, almost 50 per cent of the land area of the country is used for some type of grazing. Some desert lands have been put under irrigation to raise farm crops. Our waste lands have increased in area.

Important changes have also taken place in the volume and kinds of timber standing in the country today. Softwoods make up almost 85 per cent of the stand, the rest being hardwoods. About 70 per cent of the total stand is west of the Great Plains. Over 60 per cent of it is in the three states of Oregon, Washington, and California. Of the original volume of saw timber less than one-third remains, and of this Douglas fir makes up nearly 27 per cent, and the southern yellow pines 11 per cent. The important thing to note is that so much of our remaining stand is from two to three thousand miles from the centers of consumption in the East.

A classification of our absolute forest land shows:

Commercial forest land	462	million	acres
Non-commercial or reserved forest land	168	"	"
TOTAL	630	"	"

Of the commercial forest land 46 per cent is said to have some saw timber on it; 22 per cent of the area is covered with forests fit only for cordwood; 16 per cent is restocking fairly well for future use; and 16 per cent is restocking poorly or not at all. Non-commercial forest land is that woodland which would not pay adequate returns from a commercial standpoint, or such lands as are out of production because they have been designated as reserves or parks. Abandoned agricultural land will no doubt be added to this total to produce forest crops, as is being done in New York State and elsewhere.

BALANCING GROWTH AND CONSUMPTION

From the above figures it will be seen that about 24 per cent of our land area must be devoted to growing commercial forest crops. It has been estimated that an industrial country should have between 25 and 30 per cent of its area in forests in order to be self-sufficient and reduce imports to a minimum. Some countries have more than this. These are, in general, the wood-exporting countries like Russia, Canada, Brazil, Sweden, and Finland. The United States, besides taking care of its own needs, has usually exported from 3 to 7 per cent of its annual cut of lumber, but in recent years this trade has fallen off due to market conditions and the war. There is every reason to believe that, if we put our forest-producing lands in order, we can not only take care of our own wood-using industries, but also have sufficient lumber (especially softwood) to export.

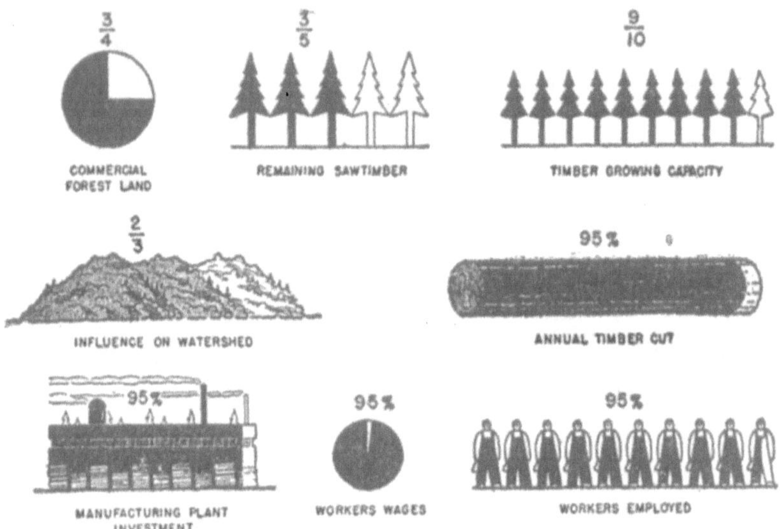

U. S. Forest Service Photo

THE EXTENT TO WHICH PRIVATE FOREST LANDS DOMINATE THE NATIONAL PICTURE

(Private lands shown by black or shaded portion of diagram)

We next come to the question: What is our present supply, our present consumption, and our annual growth? Our original stand of timber has been estimated at around 1,500 billion cubic feet, of which 519 billion are left today. This supply is being consumed at the rate of about 16 billion cubic feet per year (based upon the 1925-29 average),* indicating that our present supply would be exhausted in about thirty years if it were not for an annual growth of about 11 billion cubic feet.

The future timber requirements for the nation have been estimated by the Forest Service to be about 16 billion cubic feet per

* In 1936, considered a subnormal year, the total drain was 13½ billion cubic feet.

year. Recent war consumption has led to a revision of this estimate to 21½ billion cubic feet to take care of any increased use, new uses, export, and for a safety factor. Hence, in order to eventually balance consumption and growth, it would be necessary to increase our national timber growth up to the point of consumption. At present we are faced with the distressing facts that we are using wood 1.2 times as fast as it is growing, and we are using large-sized saw timber, which takes 100 years or more to grow, about 1.5 times as fast as it is growing. Annual drain exceeds growth by 2.2 billion cubic feet.

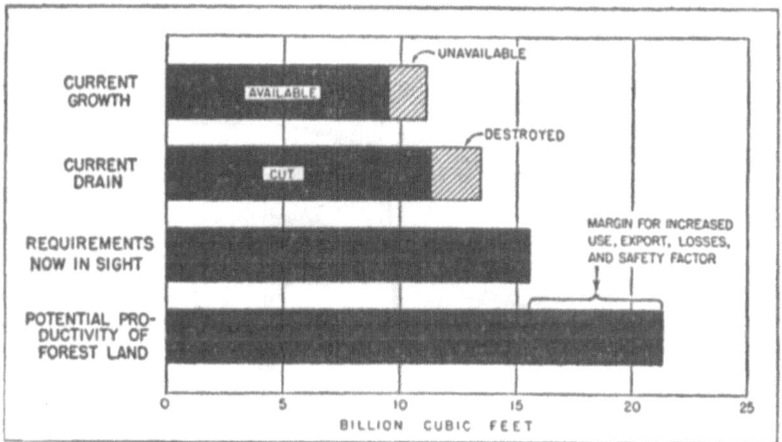

U. S. Forest Service Photo

FOREST GROWTH, DRAIN, AND REQUIREMENTS

We know in general what forestry measures should be used to secure this balance. But other vital questions arise. Can private or public ownership best do the job? How much of our forest land must be intensively managed and how much extensively? To what extent must growth per acre be increased? Can we balance our timber budget, say by the year 2000?

Public versus Private Ownership

At the present time 434,000,000 out of a total of 630,000,000 acres of absolute forest land is in private ownership. Considering commercial forests alone, 341,000,000 out of 462,000,000 acres are in private ownership. The question naturally arises: Can private ownership of 74 per cent of our commercial forest land accomplish the balancing operation and at the same time assure us of getting the other values and services rendered by the forest? Let us see.

What is the condition of our forest land today, after more than 300 years of private ownership?

Of the 462,000,000 acres of potential commercial timberland in this country, 361,000,000 acres have been cut over (in many cases also burned over), and about 77,000,000 acres of the latter are in such a devastated condition that many of them will have to be replanted in order to produce any wood crop within a reasonable time. The balance of 284,000,000 acres is so-called second growth. Only 112,000,000 acres of this contains saw timber, 101,000,000 acres have timber only of cordwood size, and 71,000,000 acres have fair to satisfactory restocking conditions.*

U. S. Forest Service Photo

But that is not the whole story. About 90 per cent of the total area of devastated and poorly-stocked forest land is privately owned. About 95 per cent of current devastation is on this same type of land. Over 95 per cent of the area burned and over 99 per cent of forest deterioration through cutting and fire occur on private land. If the past record is any indication, such figures clearly give the answer that private ownership, as constituted at present, has not solved and probably cannot solve the problem of greater timber growth for the nation as a whole. That private ownership of forest lands has failed in the past is shown also by the fact that in the Lake, northwestern,

* *Forest Lands of the United States* (House Report 323, 1941), p. 11.

and southern states over 25,000,000 acres of private forest lands are tax-delinquent to the point of foreclosure.* It is evident that if these unproductive wastes are ever to bear a commercial timber crop much of them must be acquired by the federal government, the states, or local governments where they are located.

Another phase of private ownership concerns misused and abandoned farm land. Over 50,000,000 acres of farm land east of the plains have been abandoned, and it is feared that by 1950 this area may reach 70,000,000 acres. What is to be done with this liability? Most of this land was timbered before it was cleared for farms. The land has been so depleted of fertility or eroded that it no longer grows farm crops of any value. Obviously most of it will have to be planted to forests. This is such a wide interstate problem that it will have to be done by either the states or the federal government. As a matter of fact, land adjustment of this kind has been going on since 1911 under the Weeks Law, and also since 1934 under the Agricultural Adjustment Administration and since 1937 under the Bankhead-Jones Farm Tenant Act. By 1939 about 9,000,000 acres of submarginal farm lands had been purchased by the federal government.

Another phase of the problem concerns some 465,000,000 acres of rough forest land which exerts a favorable influence in protecting watersheds. Most of this is privately owned and most of it has been cut over, and most of it is contributing to the flood and erosion problems of our large rivers. It appears as if this is also a problem for public ownership in many cases.

The record of the past clearly shows that our forest production problems cannot be solved with forest land ownership distributed as at present. National planning in the growing and management of our forest resources demands a great expansion of public acquisition in order to insure adequate total annual growth. Only in this way can our ten-billion-dollar wood industries be served with raw materials, and only in this way can our watersheds, wild-life resources, recreation, and other forest services be safeguarded. This is most potently summed up on page 41 of *A National Plan for American Forestry:*

> Laissez-faire private effort, upon which the United States has largely depended up to the present time and which is avowedly planless from the national standpoint, has seriously deteriorated

* These statements take full cognizance of the large amount of private forestry work being done, especially in the South and the far West. While highly commendable, these isolated examples are not large enough in the aggregate to materially aid in balancing the national timber budget. See last section of this chapter.

or destroyed the basic resources of timber, forage, and land almost universally. It has not concerned itself with the public welfare in protection of watersheds. It has felt little or no responsibility for the renewal of the resources on which its own industries must depend for continued existence and much less for the economic and social benefits growing out of the perpetuity of resources and industry. Even in fire protection, its most conspicuous constructive action, the public has largely carried the financial burden.

National Planning for Future Needs

What, then, should be the ownership of the 630,000,000 acres of absolute forest land so as to meet our future wood requirements? At the present time government agencies (federal, state, and local) own about 196,000,000 acres of this (National Forest areas include some non-forest land). It is thought that this figure should be increased in the near future to about 375,000,000 acres, leaving about 255,000,000 acres in industrial holdings and in farmers' woodlands. Thus we would some day have public holdings of about 60 per cent of our total forest area in contrast to 72 per cent for Switzerland, 50 per cent for Germany, 37 per cent for Italy, and 38 per cent for France.*

How are these forest lands to be managed in order to step up growth from the present average of 21 cubic feet per acre to what is necessary to meet our needs?

In order to produce the annual growth of 21½ billion cubic feet of wood needed and also to completely utilize both our commercial and non-commercial forest land, the distribution of the various types of management and their areas should be about as follows: †

COMMERCIAL:	
Intensive management	100 million acres
Extensive management	312 " "
Protection only	50 " "
NON-COMMERCIAL:	
Reserved from timber use, Chaparral and woodland to be protected, total	168 " "
Total	630 " "

Intensive management requires the best silvicultural measures to assure maximum productivity upon the most productive lands. This

* *Forest Lands of the United States* (House Report 323, 1941), p. 1992.
† *Ibid.*, pp. 1881-87.

must reach an average of from 50 to 75 cubic feet per acre per year, or roughly two to four times present growth. It requires intensive cultural methods in favored localities, like those used in Europe. Extensive management might be considered a stage intermediate between intensive management and mere protection. It would not attempt the refined cultural methods, but at the same time it would attempt to increase the growth per acre from the present 21 cubic feet to about 42 cubic feet per acre per year.

Forest areas reserved for park and other recreational use would, of course, be withdrawn from timber production, but would be protected and improved. Chaparral and woodland areas would be protected but not managed for any wood products, since their main use is watershed protection.

Some such program as the above must be put into immediate operation if it is to be fully effective seventy-five years hence, in order:

1. To get forest lands into productive use.
2. To meet our national timber requirements for forest products and services.
3. To derive the fullest economic and social benefits of the forests both for industries and the general public.

The Time Element in Forestry

Past experience in the older countries of Europe seems to indicate that forestry is an industry peculiarly suited to public enterprise and to such private corporations as are wealthy and long-lived. The "time element" in forestry business substantiates this conclusion. Whether we consider the time necessary merely to grow timber crops or to perform such forestry operations as land acquisition, fire protection, reforestation, or forest-tax reform, or the development of long-time federal, state, and private forest policies, progress is slow, and only long-time, well-conceived, sustained effort can achieve adequate results. It is right that it should be slow because of the vast interests that are involved. This calls for national planning on a large scale, with the major effort and support being made by the federal and state governments.

Federal experts tell us that forest growth cannot be increased to current requirements before the year 2000 and then only if a full and coördinated national plan, as already outlined, is carried out. It takes from 80 to 150 years to grow high-grade saw timber and it is worth remembering that about 70 per cent of our current requirement is of this size.

If we are to judge future progress by past performances, the "time element" looms up as all-important. All our progress in forestry has been slow. More rapid progress is vitally necessary and can now be safely made, because most of the foundations have been laid down. It took forty-two years to put 157,000,000 acres of National Forests under administration; twenty-two years to purchase less than 5,000,000 acres of private land; forty-five years to get about 200,000,000 acres of private land protected against fire; thirty years to get 11,400,000 acres of industrially owned forests under sustained yield management; thirty years to plant 2,000,000 acres by all agencies; and it took New York fifty years to build up a state forest domain of 2,500,000 acres. At these rates, how long would it take to put 250,000,000 additional acres under public administration, or purchase 200,000,000 acres of commercial forest land, or reforest 75,000,000 to 100,000,000 acres of poorly stocked forest land and abandoned, submarginal, farm land? Simple arithmetic gives the answers.

Let us now consider such production factors as the nature of the forest resource, absolute forest land, forest growth, rotation, capital, taxation, yields, labor, markets, profits, and stability of investment. These are the economic factors which any businessman must consider in engaging in business enterprise.

The Forest as a Replaceable Resource

Forests have so long been considered an inexhaustible resource that this American fallacy has completely warped the average man's understanding of the economics of forest production. He has, indeed, placed them in the same category with other inexhaustible resources such as land, water, and air; or he has considered them along with other exhaustible and non-restorable resources such as coal, oil, natural gas, iron, etc. The forest is in fact an exhaustible resource, but restorable under certain conditions. It is liable to great deterioration under certain methods of private use. In fact, deterioration might easily progress to the point where restoration is made impossible, as, for example, on steep mountain slopes where, after cutting, the soil is washed away, leaving nothing but bare rock.

We are, therefore, considering a commodity which is an indispensable daily need to civilized society and whose available supply is now exceedingly limited when taking into account the entire life-span of the nation and the present deteriorated condition of that resource. We are considering also a resource that can be grown on soil like a farm crop and which when ripe can be harvested like other crops and sold at a profit. We must keep in mind, however, that the forestry business, unlike farming, involves a long-time accumulation of wood capital before it is possible to harvest crops at regular intervals continuously. We do not have that wood capital today. In fact, it may be necessary to close huge areas of second-growth forests to all saw timber cutting for many decades to allow this wood capital to accumulate.

Absolute Forest Land

Land is a basic capital resource which produces wealth in the form of our daily necessities. The forest is an important factor in the economic use of land. With increased population and a higher average standard of living comes an increased need for food, wood, water, wild life, water power, recreation, and others commodities and services.

The need for greater production by our land can scarcely be questioned. The past decade has shown us that our problems are not those of overproduction but of underconsumption. American economy has by no means achieved a uniform high standard of living. Our food-producing land is not producing what it once did. We have had to import vast quantities of food. One-third of our people are reported to be undernourished. The entire rural one-third of our people earn only one-tenth of our national income.

We can adequately meet an expanding market for basic commodities by reducing waste and by making the highest possible use of every acre of land we possess. We must increase the consumer buying power of the lowest one-third of our population. We must raise more food, more wood, and more meat, in each case on lands best fitted for these respective crops. Border-line farms must be eliminated. The problem of apportioning our lands according to their potential value, so that each acre produces the crop best suited to it, is one of fundamental national importance.

In general, the lands that will be designated as absolute forest lands will be those too poor to grow agricultural crops or meat crops profitably, or which, on account of their steep slope or rocky surface, could not under any circumstances be classified as farm land. Obviously, most of this forest land lies in the various mountainous regions of the country and hence a large portion of it will be valuable not only for timber crops, but also for protection to watersheds and for wild-life and recreational use.

Farm Crops versus Forest Crops

Both agriculture and forestry must rely on natural forces—soil, air, heat, water, and sunlight—to produce crops. While the farm crop must withstand the rigors of but one or two winters, forest crops must survive the rigors of many decades. Wood crops are more hardy. They can grow in cold climates where farm crops cannot grow. Wood crops are grown largely by nature, man aiding only in a slight degree. Farm crops receive intensive cultivation and care.

The farmer produces food—starches, fats, sugars, proteins, and other nutrients—while the forester produces cellulose and wood substance. Farm crops make a great demand upon the food materials in the soil, while the forests grow well on the poorest soils. The farmer needs to add fertilizer periodically to prevent soil exhaustion. He must put back the chemicals he takes out. The forest enriches its own soil, provided the forest floor of dead leaves is left to rot. The value per acre of farm crops is, of course, greater than that of forest crops, because better soils are used. The total weight of crop, however, is in favor of the forest. Strange to say, the net income based upon the amount of capital invested may be very nearly the same for the two crops.

In forestry the growing stock and soil, and in agriculture the soil alone is the capital investment, and the crops are the interest produced by that investment. As in all business enterprises it is far more important to protect and safeguard the capital investment than it is to produce high incomes. That is the colossal mistake that has been made with reference to our national forest capital. One large forest region after another has been devastated. We have purchased prosperity and a high standard of living for some, but at the cost of depleting our basic capital resource, namely the growing stock. What basic resource is now left will not produce a wood income to meet our future needs. The same thing exactly has been done on much of our cultivated farm land. The soil has been depleted to such a degree that the final crops do not pay for the time and expense it takes to raise them. The Agriculture Department has found that 500,000 to 600,000 farms are so worn out that their occupants will literally starve to death if they continue to try to work them and that another 500,000 to 600,000 farms are nearly destroyed because of a progressive loss of fertility. Farms are abandoned because they are "worn out" and new farm lands are sought. Such a procedure, if continued, would soon lead to national disaster.

In still other ways farm and forest crops make an interesting comparison. Forests require less annual expenditure of labor, but a greater total expenditure of both labor and capital. While the farm crop matures in from one to two years, it takes forest land usually from 50 to 150 years to produce saw timber. Both agriculture and forestry are subject to the inexorable laws of supply and demand; and both are subject to taxation, although, obviously forests and forest lands must be taxed in a different manner than farm crops and farm lands. In both cases the investment must be safe before it can hope to attract capital, and both are affected by competition, transportation facilities, foreign trade, tariffs, insurance, credit facilities and similar factors.

Growth and Rotation

Two concepts peculiar to forest management need definition at this point. The first is the matter of growth—the keystone in the arch of forest production. The tree every year adds a new layer of wood over the layers of previous years, thus increasing its height, circumference, and volume. The tree crop is ready for harvest when a sufficient number of annual rings of growth have accumulated to make the tree of useful size. This may be five to ten years for bean and hop poles, fifteen to twenty years for fence posts, forty to fifty years for railroad ties, or fifty to a hundred-and-fifty years for saw-timber sizes.* This rate of growth depends among other things upon the kind of tree, the climate, the soil, and the care and method of management. Besides quantity growth, the forester also recognizes quality growth, which usually occurs in the latter years of the life of the tree. Heartwood and wood free of knots are produced at that time, and these add a great deal of money value to the final crop.

The second concept is the rotation—or the time that elapses between the first establishment of the crop and the final harvest. It may also represent the number of years which must elapse before one can return to the same area for a final crop. The length of the rotation depends among other things upon the product to be grown, the soil, climate, rate of growth of the species, and also a host of industrial, technical, and financial considerations which are unnecessary to discuss here. The rotation may vary, as has already been indicated, from five to ten years for small wood products, to over a hundred years for saw timber. The majority of our valuable timber species can be grown in from 80 to 120 years, but a few slow growing species like red spruce, white oak, and beech may require from 140 to over 200 years.

* However, if a good second growth already occupies the land these periods would be reduced.

Capital in Forest Management

Capital as a factor in forest production is of two kinds: the current or working capital necessary to carry on the business, and the fixed investment of soil and timber, which represents the capital tied up permanently and which makes continuous forest production possible. Since the labor expense for planting, protection, harvest, administration, and improvements is comparatively small—about $2 per acre per year in the German forests—the current working capital may be comparatively small, though such expenses bear compound interest from the day they are incurred to the end of the rotation.

But the most characteristic feature of the forestry business carried on for continuity is the amount and character of the permanent or fixed capital. This capital is not confined to the value of the soil (based upon its ability to produce income) but includes also the much larger value represented by the growing stock of wood which must be allowed to accumulate before the forestry business can yield an annual return in perpetuity. The money invested in both soil and wood also incurs compound interest charges to the end of the rotation.

The growing stock of a regulated forest varies with the length of the rotation, the rate of growth, and the quality of the soil. For instance, it is much greater in amount for slow-growing spruce than for fast-growing pine. Here again it is evident that large areas, large amounts of capital, and long periods of time are necessary to produce timber crops, and from the nature of the business, it is far better suited to long-lived, wealthy private corporations or public agencies to whom large investments and long periods of waiting for financial returns mean no loss or sacrifice.

Sustained Yield Management

The question naturally arises: How can a forester assure a sustained, uniform, annual or periodic yield of forest products to an industry or to a community? Much is being written about sustained-yield operation on government and private lands in the newspapers and periodicals, although its method is rarely explained. This omission is probably due to the fact that it is a highly technical subject.

A very simple case of sustained-yield management would be about as follows: Let us take a municipality that has 1,000 acres of waste land adapted to growing pine. This city could ultimately harvest a yearly crop if it would plant ten acres to pine every year for 100 years, assuming that the pine is ready to harvest at the age of 100 years. At the end of that period it could start cutting ten acres of pine a year and immediately plant it to a new crop, and keep this up indefinitely. As each crop was cut (interest) there would still remain on the ground 990 acres of forest varying in age from 1 to 99 years (wood capital). In round numbers the growing stock amounts to 98 per cent capital and 2 per cent interest, and only 1 per cent of the land is cut over annually. If we assume an average annual growth rate of 300 board feet per acre per year, the total stand would amount to 1,000 acres multiplied by 50 or the average age of the stand, multiplied by the growth of 300 board feet per year or a total of 15,000,000 board feet. The annual cut would be 10 acres multiplied by 100 years, multiplied by the annual growth or 300 board feet or a total of 300,000 board feet.

Of course, the methods of achieving sustained annual or periodic yields are numerous, but the end result is the same. In actual practice, selective cutting or group selection with natural regeneration would be followed more often than this simplified procedure of clear-cutting and planting. Also thinnings and improvement cuttings might yield intermediate returns. When periodic yields are desired, say every twenty or forty years, the crops are harvested less often over larger areas. The distribution of various age classes of timber may not be in the simple way described above. Trees of different ages may be scattered over the area in large groups, small groups, or as single trees, depending largely upon the method of silvicultural management employed to reproduce the forest. In any event only the growth for a given year or a given period of years would be cut, the wood capital remaining intact.*

Taxation of Forest Lands

One of the recurring annual expenses which the forest owner must meet with considerable trepidation is taxes. In the past, forests have been taxed like any other real property. Timberlands are appraised every year and taxed according to their value, irrespective of whether they are yielding any returns. Since the theory of taxation is that it must be based on ability to pay, it is difficult to see how such

* Without going too deeply into forest finances, it might be pointed out that there are numerous aspects of the compound interest calculations in forestry. W. Schlich, in *Forest Management* (London, 1905), states that in determining the value of a forest property all calculations must be made with compound interest, since all money, whether principal or interest, is capable of again yielding interest. In determining the expectation value or what can be done, the true aspect is discount. A forest producing annual crops and therefore annual financial returns would not incur compound interest charges.

an antiquated property tax could encourage forest production. As a matter of fact, it is pretty well agreed that excessive taxation of forest land has been a large factor in liquidating existing virgin timber and in preventing the land's being developed for continuous forest crops. Foresters and lumbermen are coöperating to change this.

If taxes are to be based on the income-producing value of property, forests should be taxed only when cut, and the land should be subjected to a nominal yearly tax based upon its actual value. The history of forest taxation shows a gradual trend towards this principle, although with considerable variation in the method by which it is done. Most of our important forestry states have improved laws aimed to encourage timber growing.

About the time of the Civil War, in order to encourage tree planting, laws were passed in some of the prairie states granting tax exemption to tree planters or giving bounties to these settlers, or giving rebates on tax payments. About 1911 began the era of yield-tax laws, and up to 1932 some seventeen states had passed such laws. They varied a good deal in effectiveness. The best of them, passed in some of the New England states, made growing timber exempt, placed an ad valorem tax on the bare land, and collected a yield-tax or a personal property tax when the timber was cut. A few states have abandoned the ad valorem tax on the land and now have a flat-rate tax per acre of timberland, together with a yield-tax at harvest time.

Obviously a great drawback to the yield-tax or any other reduced or delayed tax is that it may cause great fluctuation in the annual income from taxes, especially in counties where much of the assessed property is in the form of timberland. Michigan and Wisconsin in recent laws call for a state contribution to local communities in consideration of the loss of local revenue due to the modification of the property tax, this advance to be repaid when the timber is cut and the yield-tax is applied. Most of the recent laws call for the special classification of forest lands to be given preferential treatment or exemption from taxation.

Labor in Forestry

In certain European countries where forestry and agriculture have developed with equal vigor side by side, we find that agriculture uses from ten to thirty times as much labor as forestry. Farming is labor-intensive; forestry is labor-extensive. The farmer plows, harrows, and cultivates his crops and even sprays his crops against insects and diseases. The forester does none of these things. The fact that most forestry work can be done in the fall and winter months and most farm work is done in the spring and summer months makes one industry supplement the other. This results in less unemployment, an important factor in a densely populated country.

Labor in the forest is technically simple; very little skilled labor is needed; and not much organization and specialization is needed once the enterprise is established. Unskilled labor is usually abundant and cheap. While one man could manage a farm of 250 acres, it would take about 25,000 acres of forest land to keep one manager fully occupied. Of course, both managers would require much temporary help. On the National Forests there is today one district ranger for about 200,000 acres and a fire guard or assistant ranger for every 25,000 acres.

The technical skill and foresight needed for forestry enterprises is greater than for farming. The forester in formulating working plans over long periods of time for his forestry property requires a good deal more technical knowledge and a great deal more judgment and foresight than the farmer, who can change his methods and his crops from year to year to meet the new conditions. An error in judgment on the part of the farmer rarely means the financial loss of more than a year or two; on the part of the forester it may mean considerable loss of time and money, often extending over decades.

As pointed out elsewhere, if one of the broad national plans for forest management recommended in the Copeland report were adopted it would mean the employment of hundreds of thousands of rural workers in forestry work. And these jobs would be available year after year. The employment would generally be at off seasons for farm labor. Furthermore, forestry can delay certain operations for years if need arises (poor market conditions or a scarcity of labor), and it can give that labor employment when it needs it most—in times of economic stagnation.

Markets and Wood Prices

The question is often asked: Is it really going to pay us to grow our own wood supply? And this question naturally suggests several others: Does not the decreasing consumption of wood in this country in the last twenty-five years indicate that modern civilization is going to use less wood in the future? Is not wood being replaced by competing materials? What is the relation of world consumption to available world supply? If other countries have an abundant supply can we not import our necessary wood from these countries?

It is pretty well agreed that the great lull in the world timber trade, and especially the sharp decline in consumption in the United States from 1930-40, were both due in a large measure to disturbed economic conditions. The Forest Service says that we can look for a future demand at least equal to the 1925-29 level of about 16 billion cubic feet per annum. Our own consumption is due for a sharp rise above the 1930-40 level when economic conditions warrant it. We are told that we are short about 12,000,000 dwelling houses.* Not only must this deficit be made up, but the normal building of from 300,000 to 400,000 houses annually must be resumed. About one-third of our citizens are said to live in sub-standard houses; over 60 per cent are said to be in need of some kind of repair. All of which seems to indicate that the drain on our forests will continue, since wood is still the cheapest building material known. Great Britain in a similar way is looking for a steady demand at about the level of recent normal years, to say nothing about the reconstruction that must follow the present war.

Competing materials seem to have had little effect on the world consumption of wood in the last fifty years, for the uses of wood are constantly changing. As old uses are eliminated by competing materials, new uses arise. As wood uses in simple carpentry get less important, the use of wood cellulose for paper, cardboard, and plastics is rapidly increasing. Experience seems to indicate that the modern industrial and agricultural world cannot get along without wood. As a matter of fact, the world need of wood, especially softwoods, is increasing as industrial expansion spreads to agricultural countries. Finally, normal world timber consumption far exceeds present growth, and the world supply of valuable softwoods is steadily diminishing.

What is the outlook for getting foreign timber in the next 100 years? Wood is a world commodity and as such will always be exported from countries which have an abundant supply to those that have a limited supply. Is this possible today? About 80 per cent of our lumber consumption is of softwood timber. The United States, Canada, and Russia contain 80 per cent of the world's softwood supply. What are the conditions in these countries? In both this country and Canada, cutting, together with losses from fire, insects, and diseases, far exceeds the annual growth. Increased cutting in Russia, accompanying the present vast industrial expansion in that country, holds out little hope for large supplies from that source.

*National Housing Administrator estimates the shortage as 12,600,000 dwelling units to make up the accumulated deficit in building, provide for returning service men, and replace substandard houses. *New York Times,* October 22, 1944.

There seems to be but one answer: If we are to have a cheap and abundant wood supply, we must grow it ourselves.

A diversified wood market and higher wood prices are natural concomitants to continuous forest production. As long as we use virgin supplies that cost us nothing to grow, wood prices will be comparatively low. The expense of growing timber is considerable and that, of course, will be reflected in higher prices. A permanent forest industry, on a continuous production basis, harvests wood of all sizes. Thinnings and improvement cuttings made every ten or twenty years during a rotation would throw on the market a most diversified assortment of forest products. In Europe often from 30 to 40 per cent of the total yield is in the form of these intermediate cuttings. We must have markets to absorb this material, consisting of small poles, posts, railroad ties, large poles, firewood, piling, and other commodities. It is very likely that in the future the complete utilization of a large forest tract will call for not only a lumber mill but also pulp mills, excelsior mills, veneer mills, toy manufacturing, and other industries to use the odds and ends of lumber manufacturing. The higher wood prices go, the more complete must be our utilization.

Safety of Investment

In a region well developed industrially and agriculturally to insure good markets, where reasonable protection against fire and insects is assured, forestry is a safe, long-time investment. If damage by these agencies can be kept within narrow limits, the loss can be treated in the nature of a fixed annual charge. A forest, once established and placed under management, yields relatively uniform annual returns, while farm crops vary greatly from season to season. Temporary high wood prices can be taken advantage of by cutting more than the normal yield, and low prices can be circumvented by omitting cutting. That is, the prices of forest products are subject to less sudden fluctuation than the value of money, measured by the commodities it will buy. Also the forest soil itself offers practically absolute safety since under proper management it is indestructible. Finally, forestry offers permanent and continued investment of large capital sums and thereby avoids the losses due to repeated reinvestment. Thus fluctuations due to business cycles are largely overcome.

Profit or Rate of Interest Earned

The rate of interest earned by any industry depends principally upon the security of the investment and the certainty of its yielding a return, upon the supply and demand of capital in general, and upon the general credit conditions in the country. In general, the more secure the investment is, the lower the interest rate accepted by investors. Probably the best indication of the general credit conditions of a country is the interest rate yielded by the government bonds. Of course these conditions may change from one locality to another and from one year to another. Hence it is evident that the interest rate applicable to forestry cannot be fixed with certainty for any given locality or for any given period, since it depends upon a variety of technical and economic factors. Rather must we think of the rate of interest earned in forestry in terms of average financial conditions over a long period of time.

Some idea of what this interest rate should be can be secured from European experience before 1914. As already stated, the interest rate earned by large forestry enterprises (also the rate to which the forestry business is entitled in case it wants to borrow by issuing securities) must be considered in connection with the nature of the investment. Forestry is considered in Europe a safe, long-time investment and as such the interest rate earned is comparatively low compared to some of the big profits that are yielded by less secure short-time enterprises. Germany, before the First World War, had 35,000,000 acres of forest with a total capitalization of over $6,000,000,000 or an average of about $180 per acre ($25 for the soil and $155 for the timber). This property had been for years producing a constant revenue of 3 per cent on the capital investment, yielding a continuous annual income of about $190,000,000 a year. (In 1940 our own 175,000,000 acres of National Forests yielded about $5,860,000). But it must be remembered that the value of German forests is constantly increasing and hence the income is increasing. And above all, it must not be overlooked that this income is from land too poor to grow agricultural crops!

The objection is sometimes made that a 3 per cent return would not be very attractive to investors. Yet agriculture, one of the country's basic industries, over a long period of years, taking the good years with the bad, has done no better than 2 to 2½ per cent. At the present market a relatively safe, long-time investment would not net over 3 to 4 per cent. Many long-term government, state, and municipal bonds are being floated at the present time for much less than these rates. Most of the New Deal financing between 1930 and 1940 was at a rate between 2 and 3½ per cent.

As to forestry investments in this country, the meagre data available seem to indicate that our most rapid-growing species on the better soils will yield between 3 and 5 per cent on the investment, and this does not take into account increased wood prices, which must come about in a country of growing timber scarcity, nor increase in soil values, nor increases in soil productivity, factors which will very likely make forestry a very attractive business in favored localities.

Other factors also must be taken into account in appraising the possibilities of the forestry business. A number of our tree species will do better than those of Europe. We have a better climate, better soils, and a greater variety of trees to choose from. Whereas large forest administrations in Europe produce on the average of 55 cubic feet of timber per acre per year (about 350 board feet) we have large favorable areas in the South and West that probably can produce from 50 to 75 cubic feet per acre per year under intensive management. Several valuable fast-growing species on the best forest soils, like Douglas fir on the coast, white and red pine in the Lake states, and loblolly and other pines in the South, will probably produce well over 100 cubic feet of timber per acre per year in short rotations. This, of course, is a striking contrast to the present average annual growth, which is only 21 cubic feet per acre per year.

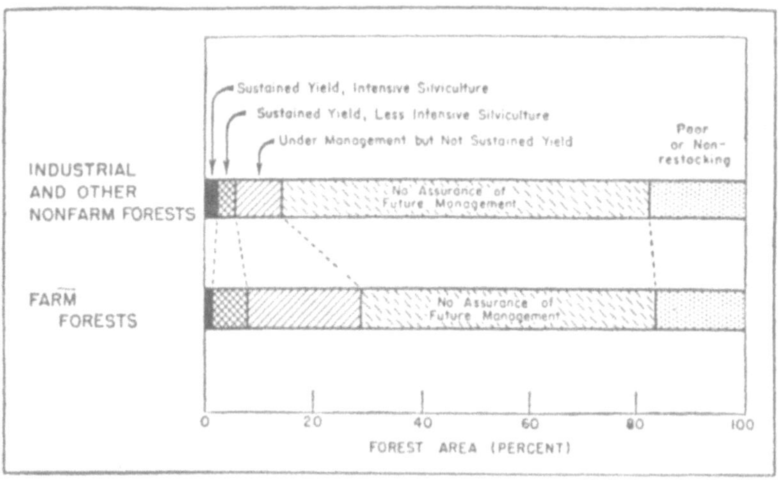

U. S. Forest Service Photo

CHARACTER OF FOREST MANAGEMENT
Private Lands

Present Extent of Private Forestry Practice

The extent to which the economic factors of forest production just enumerated are considered favorable for engaging in the practice of private forestry is reflected in the number of acres of forest land actually placed under management to date. The latest summary of this practice both on woodlands owned by farmers and on industrial holdings is given in the following table:

Table 9

STATUS OF FOREST PRACTICE ON COMMERCIAL PRIVATE LANDS, 1938 *

Types of Management	Commercial Farm Woodlots		Industrial Lands	
	Area, million acres	Per cent of Total	Area, million acres	Per cent of Total
Total	138.8	100.0	202.1	100.0
Intensive sustained yield	1.6	1.2	3.9	2.0
Extensive sustained yield	9.7	7.0	7.5	3.7
Extensive (not sustained)	30.0	21.6	17.2	8.5
Partially productive	76.2	54.8	136.8	67.6
Lands poor to non-restocking	21.5	15.4	36.7	18.2
Fire Protection Status:				
With fire protection	93.0	67.0	141.0	70.0
Without fire protection	46.0	33.0	61.0	30.0

* *Forest Lands of the United States* (House Report 323, 1941), pp. 22 and 25.

This table shows that about 30 per cent of the commercial woodlots and over 14 per cent of the industrial holdings are under some form of management. Taking all these holdings together, the table reveals that over 80 * per cent of 341,000,000 acres of commercial forest land is in a productive or partially productive condition and that less than 20 per cent will need artificial restocking. The table shows also that about one-third of these lands are still without fire protection.

It should be noted that over 11,000,000 acres of industrial timber holdings are being managed under some form of sustained yield.

* A recent survey by the National Lumber Manufacturers' Association of private forest land in 36 states shows 89 per cent in productive condition. Such statements have little significance, however, unless we know something of the age, quality, species, density, and accessibility of forests on the productive areas.

Such holdings are mostly in the Pacific Coast, southern, and New England states. The most noteworthy private attempt at providing a future timber supply is a movement known as "tree farms." These are small or large areas set aside by lumbermen, timberland owners, and farmers dedicated to growing forest crops for commercial purposes. They are protected and managed for continuous production in the belief that forest production is an obligation of forest owners and of the industries using forest products. The National Lumber Manufacturers Association, working through its regional associates, such as the West Coast Lumbermen's Association, has sponsored about 8,000,000 acres of tree farms in Washington, Oregon, Arkansas, and Alabama. The owner of the tree farm agrees to use the land for the production of forest crops; provide reasonable protection from fire, insects, diseases, and grazing; cut the forest crop in a way that will insure future cuts; and furnish the association with reports of progress on his tree farms when called upon.

Many cases of sustained-yield management could be cited, but space permits the mention of only a few. Certain lumber companies operating in some of the towns of the Cascades of Oregon had the choice of cutting all and getting out, thus causing the towns to be abandoned, or gathering up enough timber, both public and private, to cut on a sustained-yield basis. They chose the latter course, cutting only a certain percentage of the mature timber and returning periodically for another cut.

The Crosett Lumber Company with 400,000 acres in Arkansas and Louisiana is treating its pine and hardwood forests the same way by going through the stands once every fifteen years for a cutting. Close utilization of all products required the building of a chemical distillation plant and a pulp mill.

The California Fruit Exchange Association, which uses huge quantities of lumber for fruit boxes, has for over 20 years cut its own and National Forest holdings on a sustained-yield basis and is operating a very modern mill at Susanville. The over-mature ponderosa pine forests are gradually being transformed into rapidly growing, productive stands and forest fires have been kept definitely in check.

The localities apparently most favorable for this forestry practice are the South, Pacific Coast, New England, and the middle Atlantic states. Then follow in order the central, northern Rocky Mountain and the Lake states. The species of trees most favored on account of their intrinsic value and their rapid growth were loblolly, shortleaf, and slash pines in the South; redwood, sugar pine, and Douglas fir on the coast; white pine and spruce in New England; and western white pine in the northern Rocky Mountains.

Measures Needed to Aid Private Forestry

Under our system of free private enterprise, foresters, lumbermen, and public would like to see private forestry work out its own problems without government supervision. However, it is pretty well agreed that in changing over from a liquidation policy to one of good forest management, the private owner is confronted by many problems for the solving of which public aid is imperative. Such public assistance forthwith leads to more or less government supervision and control.

Some of the pressing problems facing the private owner * are: (1) fire, insect, and disease damage; (2) uncertain markets due to competing materials; (3) imports of foreign forest products; (4) the tangled ownership problem; (5) oversized plants and excess mill capacities; (6) excessive investments; (7) high property taxes; (8) unsound financing; and (9) other financial burdens incident to bringing deteriorated lands to a productive condition. In too many cases such unsolved problems are leading to rapid liquidation, land deterioration, and the attendant economic and social evils.

The Forest Service has estimated † that the future distribution of ownership between public and private forests should be 39 per cent private and 61 per cent public. This would mean a reduction of 200,000,000 acres from the present total of 434,000,000 owned privately. By a gradual process of elimination only the most favorably situated forest lands will remain in private ownership, principally those located near good markets, those having good growing conditions, and those with growing stocks not depleted by fire, overcutting, and other factors.

The probable future development of private forestry may be stated in three major steps: (1) the passage of fundamental fire protection and taxation laws to promote free private enterprise; (2) failing to get sufficiently good results in this, public financial and management aid which naturally brings with it restrictions and more or less public supervision; and (3) where public supervision fails, especially in areas where the public interest is vital, the last alternative of public ownership must be used. At the present time the principal measures necessary for the acceleration of private forestry practice come under the first two steps. The following program is designed to assist the

* *Forest Lands of the United States* (House Report 323, 1941), p. 24.
† National Resources Board *Report* (1934), p. 21.

private owner in solving the nine pressing problems enumerated: *

FOREST PROTECTION: Fire protection should cover all forest lands. There should be increased appropriations under the Clarke-McNary Law providing funds to states and there should be an extension of the same law to cover coöperative protection against insects and tree diseases.

FOREST TAXATION: Some form of equitable taxation of forest land will have to be evolved to lessen the burden both on immature and mature timber. A low land tax levied annually while the timber is maturing and a severance tax when the timber is cut seems to be the best plan. Local government units should be compensated yearly by the state or federal government for loss of revenue by means of revolving loans to be repaid when severance taxes are levied. This general principle will have to be modified to fit local political, economic, and forest conditions.

FOREST CREDITS: Even with the above two difficulties removed, many large private owners would still be handicapped financially in starting a continuous forestry enterprise. There should be available long-term, low-interest federal loans to private operators through existing federal credit corporations to stop the liquidation of forest enterprises, consolidate holdings for more efficient management, provide improvements, and in other ways finance the transition to sustained-yield management and the readjustments needed in the lumber industry itself. The loans are to be granted under the provision that the forest land be kept in a continuous productive condition and under other necessary safeguards to the lender.

FOREST ORGANIZATION: The consolidation and stabilizing of forest ownership into natural economic units by purchase or exchange; building up the growing stock for future needs; and the liquidation of the over-mature timber as fast as the market conditions will warrant. Some of this organization will have to be done with the aid of loans.

OTHER PUBLIC AID: Appropriations through existing laws for reforestation, forest extension, forest research, and other work, and through new legislation for farm forest coöperatives and for coöperative sustained-yield units and coöperative management agreements with private individuals or corporations.

* *A National Plan for American Forestry* (Sen. Doc. 12, 73rd Cong., 1st Sess., Washington, D. C., 1933), pp. 271-314; National Resources Board *Report*, pp. 206-16; *Forest Lands of the United States* (House Report 323, 1941), pp. 24-33; G. S. Wehrwein, *Land Economics* (New York, 1940), pp. 271-314.

An economic movement cannot get ahead of the people who are concerned. Unless the rank and file are educated to the purposes and results of good forestry, little progress can be made. This is especially true in fire protection. A new land-use concept must come into the public consciousness. Forests have been taken for granted as a place from which to get firewood and lumber *ad infinitum.* Now both the forest owner and the public must completely about-face. Both must realize that forests, like farm crops, can be grown profitably as an annual crop if the owner makes some present sacrifices.

More important still, forest lands even in private hands must be administered as a trust belonging to the people. A study of the many services that the forest renders actually makes the forest a public utility. Forest land must continuously grow crops no matter who owns it. For economical management such lands should be consolidated and not scattered, as it often is now, in checker-board fashion. For the best results such sustained-yield units must be preserved intact generation after generation if held in private ownership. The death of the owner should have no effect upon the continuity of management.

Public Regulation of Private Forest Lands

The term "public regulation" seems to have acquired an unpleasant connotation, and to many forest owners still implies government interference. As already suggested, government help of any kind almost always carries with it certain obligatory conditions under which the aid is granted. As long as the private owner and the states cannot effectively handle the forest resource problem, public regulation is inevitable. Let us examine the trend of recent thinking on the subjects and also inquire into public regulation in other countries.

The vast extent to which forest land enters into our national economic and social structure has been clearly indicated in the last few decades by thousands of pages of printed matter, by the spoken convictions of the men on the ground (foresters, lumbermen, farmers, industrialists, sportsmen, etc.), and by the trend of curative proposals and forestry legislation. Among the more significant developments should be mentioned:

1. A series of highly important coöperative forestry laws enacted by Congress since 1911 which have greatly improved the general forestry situation.

2. Many state forestry laws designed to aid the practice of private forestry, especially since 1925.

3. The report, *A National Plan for American Forestry*, issued in 1933, which brought to light many challenging facts.

4. Reports of the National Resources Board (1934) and its successor the National Resources Committee (1935-39) emphasizing the place of forestry in the whole conservation picture.

5. Hearings before the Joint Congressional Committee on Forestry, 1938-1941, the publication of the report, and the subsequent ameliorative legislation placed before Congress. Here, for the first time, was assembled the experiences and opinions of the men in the "front line" of the forest conservation fight.

These laws and reports indicate a definite trend toward the public regulation of private forest lands. Post-war legislation will doubtless give the trend definite form and direction. From the numerous reports we cannot escape the conclusion that there exists on the part of the private owner, the public, and the public's representatives in government a heavy responsibility for safeguarding the public's interest in forest resources, regardless of who owns the forest land, and that concerted action and early remedial legislation are needed. The rôle of wood as a strategic war material has driven the lesson home.

The responsibility is particularly heavy on the private owners, since they hold three-fourths of the best forest land and the most accessible three-fifths of the saw timber, and at the present time produce 95 per cent of the total timber cut.* Also, a very definite obligation rests with the private owner, who, in spite of his destructive policies, is yearly benefitting by the expenditure of millions of public funds for coöperative fire protection, forest research, forest restoration, and for erosion and flood control, to say nothing of the large public relief expenditures recently made where forests have been exploited to the vanishing point.

The private owner is also obligated to use forest land as a permanent resource—a public utility, as it were—so that it may contribute its share of security and stability to forest-dependent families and communities, and so that it may eventually contribute permanently to regional and national welfare and to national defense. Such permanent forest use embraces not only timber and other forest products, but also such services as erosion control, watershed protection, recreation, wild-life restoration, and forage.

Forest land also involves a vital public interest. Too long has the public sanctioned forest exploitation and the economic and social ills which are part and parcel of it. It has stood idly by and allowed devastation to progress to such a point that only public action can

* *Forest Lands of the United States* (House Report 323, 1941), pp. 26-28.

save the resources and the industries. To that extent the public can be justly blamed for our present debacle.

The public also has a number of other responsibilities. It must see to it that governmental units enact legislation that will lift the burdens from private forestry practice. In a democracy the people make the laws through their chosen representatives. For example, they must take more aggressive action to remove the inequalities of the unmodified property tax. The public is also responsible for adequate forest fire protection in the form of voting more adequate funds. It must also take almost full—90 per cent—responsibility for the careless setting of fire in the woods. It must see to it that more public funds are used to support research, the fountain-head from which all forestry progress flows. Finally, simple justice and practical necessity demand that the public, through various governmental units, take over and manage a substantial part of the forest lands now in private hands, especially of those owners who cannot or will not be aided by coöperative laws or those that cannot be persuaded to practice forestry on a voluntary basis.

A number of regulations have already been passed by the various states dealing with protection, reforestation, taxation, and other phases of management.* All states make it a misdemeanor to set fire to the forest; thirty-nine states provide controls for camp fires; forty-one states have special requirements for the use of mechanical equipment in the forest; thirty states have slash and brush-burning requirements; ten states place the duty of forest fire protection on the private land owner while thirty-eight recognize such protection to be the duty of the state; forty-one states have enacted insect and disease control legislation; four states require that seed trees be left after logging; three states set diameter limits for cutting; and four states require forestation of private lands after lumber operations. Legislation also deals with related problems: forty-two states have timber trespass laws; eighteen states encourage reforestation by totally or partially exempting the land or the forest crop from taxation, at least until harvested; and four states provide deferred taxation on mature timber.

No attempt has been made to require that the handling of the forest shall not entail damage to others through erosion or the irregular flow of streams. Nor in general, have the states attempted to prevent the owner from damaging or destroying his forest; nor to

* *A National Plan for American Forestry* (Sen. Doc. 12, 73rd Cong., 1st Sess., Washington, D. C., 1933), p. 1050; *Forest Lands of the United States* (House Report 323, 1941), p. 1744.

require him to keep his land continually productive; nor to preserve recreational and wild-life values for the benefit of the public.

In Europe * restrictive measures dealing with forest management were first aimed at the protection of watersheds and later at providing an adequate national timber supply. The widespread devastation of mountain forests that followed the period of the French Revolution resulted in disastrous floods, with great damage and loss of life. Hence, immediately after this period most of the restrictive measures were aimed at preserving the protective function of forests rather than at preventing a timber shortage. Since 1900 the scope of public control has been increased and broadened. The maintenance of an adequate timber supply has been a prominent factor. The growing spirit of nationalism, designed to make each country self-sufficient, was a powerful incentive in the period between the two world wars.

World-wide experience has shown that uncontrolled private ownership has always led to devastation. In practically every civilized country in the world, the accepted solution has been some form of state intervention and state control of forests.†

A recent survey ‡ of the forest regulatory measures of forty-six foreign countries has summarized the following generally accepted principles of government supervision, which are modified in this or that detail to suit conditions in various countries:

1. The owner is free to manage and utilize his forest so long as he does not directly or indirectly injure others or the public welfare.

2. The maintenance of a continuous forest cover to insure the protective function of forests is mandatory, because the destruction or mismanagement of forests which serve to hold the soil in place, conserve water, regulate stream flow, protect public health, and promote the national defense is certain to cause injury to others.

3. Protection forests are classified as such by the proper authorities after appropriate investigations and hearings.

4. Cutting in classified protection forests may be done only with the permission of competent authorities. The methods used are outlined in detail and reforestation is compulsory.

5. The object of control over other than protection forests is merely to keep the land in a productive condition. Misuse destroys productivity and must be avoided. Sustained-yield management is usually not required. Natural or artificial reforestation is required,

* *A National Plan for American Forestry* (Sen. Doc. 12, 72nd Cong., 1st Sess., Washington, D. C., 1933), p. 1005.

† W. E. Hiley, *The Economics of Forestry,* New York, 1930.

‡ *A National Plan for American Forestry* (Sen. Doc. 12, 72nd Cong., 1st Sess., Washington, D. C., 1933), p. 1037.

but there is no attempt to compel owners to grow any particular kind or quantity of material.

6. Public control is usually decentralized under the local or provisional boards or commissioners composed of forest owners, technicians, administrative officials, and local inhabitants. It is truly on a democratic basis.

That summarizes the hundred years of experience of fifty of the major forestry nations of the world, and such regulations might well serve as the basis for the control of private forests in America, modified, of course, to meet the political, economic, and financial conditions in the various forest regions of the country.

Federal versus State Regulation

As far as possible, private enterprise in forest industry should be allowed to work out its own problems, aided by the combined forces of education, economic encouragement, and far-sighted public and private coöperation. As just pointed out, however, no country has solved its forest problems completely without some measure of public regulation. Certain simple restrictions on timber cutting and forest land management are both necessary and desirable. The combined experience of other nations shows this. The big forestry question of the moment is whether this shall be by federal or by state governments.

Private forest industry favors state regulation,* because states are more closely in touch with local conditions. In the words of Dr. Compton, Secretary-Manager of the National Lumber Manufacturers Association, these regulations "should be based on the tested experience of forest land-owners in each region; they should be locally administered under state laws; and state forestry commissions competent to administer such state forest practice codes should be built up in each state." The apprehension is often expressed that a federal bureau will be set up to dominate the lumber industry; that federal regulation would tend to nationalize all forest resources; and that eventually this would lead to the substitution of public ownership for private ownership, and the development of state socialism.

The Forest Service, on the other hand, claims that federal regulation is necessary to secure the proper handling of forest resources because state regulations unsupported by strong federal legislation and financial help cannot be effective and because federal financial aid to states may not induce all states to enact appropriate legislation and follow a uniform and satisfactory standard of forestry practice.

* *American Forests,* Nov., 1943, p. 548.

It is contended that the Secretary of Agriculture should have authority to set standards for required forest practice; to pass on practices adopted by states; to inspect enforcement; and to take direct action where enforcement and standards are not acceptable.

These two viewpoints are certainly not impossible of compromise. Since continuous forest production is the end sought by both (and the very life of the lumber industry depends upon whether this is achieved), it seems that local legislation and enforcement with a minimum of federal supervision would achieve the ends desired. In fact, it is hard to believe that the lumber industry would not adopt any reasonable measures to insure its own existence. There is ample evidence that this is the case. The lumber associations of the Pacific Coast and their operating companies are to a large degree responsible for improved tax and fire legislation in these states and for the widespread campaigns to cut down the forest fire hazards. This improved background for continuous forest production is an outstanding example of progressive state-private coöperation.

The Development of Permanent Forest Industries in the Douglas Fir Region

The large forest industries of the Pacific Coast are today in the process of evolution from the old-fashioned "cut and run" basis to the modern "harvest and stay" economy.* This section of the country has made most impressive progress in the last twenty-five years. The past with its errors and omissions is a closed book and the attention of lumbermen and trade organizations alike is focused upon the future. Most of the forest industries are today committed to efficient fire protection, and 70 per cent of them are practicing conservative logging, leaving adequate seed trees, and in other ways providing for a continuous yield of forest products. Forest nurseries are also providing planting stock.

Of course, this does not mean that most lumbermen are practicing forestry. A sufficient number of them, however, are sold to the idea of keeping their forest lands continually productive, so that it is no exaggeration to say that the Pacific Coast is becoming forest conscious. Furthermore, this swing to scientific forestry principles means that the courage, vision, ingenuity, and enterprise that made the lumber industry a great constructive force in our development as a

* Based on *We Are Growing Trees, Timber Is a Crop,* and *After the War, Wood,* by Roderic Olzendam, Director of Public Relations, Weyerhaeuser Timber Company, Tacoma, Wash., and *A Decade of Progress in the Douglas Fir Region,* published jointly by the West Coast Lumbermen's Association and the Pacific Coast Loggers Association, Seattle, 1943.

nation, is now being harnessed to make forest industries a creative, expanding, and progressive force in American life. As the old mature growth becomes exhausted, it is more and more evident that the young thrifty stands of second-growth will have to be protected and managed to insure a future cut. Both individual lumbermen and their trade organizations realize this.

The Douglas fir region of the West Coast is a timber region of great potentialities. There are about 26,000,000 acres of commercial forest land in the states of Washington and Oregon, of which the public owns about 45 per cent. The total stand is 550 billion board feet, which is over one-third of the nation's entire supply. Climate and soil combine to make this one of the great forest-growing sections of the nation. Social, economic, and industrial conditions also are favorable for timber production.

This area is supplying a large part of our present timber needs and is now being cut at the rate of eight to nine billion feet per year. The total growth, however, is estimated at only about two-thirds of this amount. In 1933 the Forest Survey showed this area to contain: old growth, 39 per cent; growing stands over 16 inches, 17 per cent; growing stands from 6 to 16 inches, 18 per cent; restocking areas under 6 inches in diameter, 11 per cent; cut since 1930, 6 per cent; and areas less than ten-per-cent-stocked, 9 per cent. A resurvey in 1940 showed that although about 700,000 acres of stagnant mature forests were logged during the seven-year period, the restocking areas gained over one million acres. At this rate of young forests replacing old stagnant stands, the entire region can look forward to the time when total growth will catch up with total depletion and the region will be upon a sustained-yield basis. It appears that it will take about forty years for the timber budget of the region to become balanced, provided the various factors which enter into the problem continue as they are at present. Of course, better fire protection, better management, and reforestation will shorten this period.

Such large areas of absolute forest land leave no other alternative. They are not agricultural; they must grow forests. Furthermore, great amounts of capital are invested in sawmills, logging machinery, and forest lands and the industry directly supports half-a-million people. Approximately $6,800 invested in timber, tools, equipment, railroads, rolling stock, and machinery are needed to employ every one of the 130,000 workers. In some areas more than 60 per cent of all the industrial wage earners are associated with forest industries, and these workers pay from 10 to 50 per cent of the total taxes which provide social benefits for all the people. In many parts of the West

75 per cent of all the outbound water freight and 60 per cent of the outbound rail freight originates with the forest. Looking at the industrial picture from another angle, the 1937 cut of six-and-one-half billion feet was valued at $15,000,000; yet the total public returns in taxes, direct payrolls, supplies, depreciation, interest, and freight revenues was well over $200,000,000.

To this impressive total must be added many conversion industries such as furniture, fuel, pulp and paper, shipping containers, plastics, chemicals, etc., which use forest materials and which also produce employment, wages, and money to run local governments. Lastly, forage, recreation, wild life, and water supply must be listed as important forest services.

Pacific forest industries are awake to the needs of the situation. There will be timber to cut after the virgin supplies are gone. As these old stagnant forests are cut, many of them making no net growth at all, they will be be replaced by rapidly-growing, thrifty, young stands. More complete fire protection will save thousands of acres of young growth for the future. There are now almost 200,000,000 acres of potential "true farms" in eleven western states and British Columbia, and about 2,500,000 of these have already been registered by their owners. Forest industries have established tree nurseries to raise stock to supplement the seedlings coming up naturally from seed trees left by loggers. The large forest fire associations have spent $10,000,000 in the last ten years in Washington and Oregon alone for fire prevention, detection, and suppression. And it is stated on good authority that, if we were to look for examples of good forest management in the West, we would find it on private land quite as often as upon government-owned forest land. In this connection it must also be pointed out that the industry is employing more and more technically trained foresters, about 240 being now employed in management, research, manufacture, and utilization.

In western Washington the progressive loggers and lumbermen of the Douglas fir region have developed a forest-industry tree nursery producing about 5,000,000 seedlings a year for planting on non-restocking burned and cut-over areas. It covers about forty acres and uses many modern and ingenious methods for propagating young forest trees. It is practically completely mechanized. Member companies purchase their stock from this central nursery.

But private forest industries have been active in other ways. They have to their credit an educational program to make forest visitors forest-conscious; to apprise the public of the true status of our national supply; to assist people in the best uses of wood products; and above all to get Americans to "stop looking backwards and look ahead without prejudice." The industry has helped solve the vexing forest taxation problem and by its "Keep Oregon Green," "Keep Washington Green," and "Keep America Green" campaigns has brought the fire menace to the very door of the individual forest user.

Thus fire and taxes are gradually being eliminated as deterrent factors in timber production. Yet the uncertainty of the federal government's attitude toward private enterprise in general and private timberland in particular is as much a retarding factor as fire and taxes were decades ago. The great increase in the centralization of powers in Washington, D. C., even before the war, accompanied by many bureaus and commissions promulgating rules and regulations without Congressional approval is considered by many to be a menace to free enterprise. As far as the lumbermen are concerned, these rules run the gamut from foreign trade all the way to lumber-selling practices and they make any business man wonder what is coming next. New bills dealing with regulation of private forest lands pending in Congress add to the uncertainty. In 1933 the government approved a good Forest Industries Code (N.I.R.A.); yet less than two years later it was declared unconstitutional and abandoned. Sound forestry practice cannot be built upon such unstable conditions. The industry as a whole wants to operate under the system of free enterprise, and it maintains that it can flourish only if motivated by encouragement and coöperation rather than by bureaucratic coercion.

Now let us look at the question as to whether forestry will actually pay a profit in this important Douglas fir region. The best available information, based on Forest Service yield tables,* seems to show that most privately owned forest land can produce between 500 and 1,200 board feet per acre per year, and computations show adequate returns over cost to warrant holding, protecting, and restocking timber lands.

On most areas the management of Douglas fir is a comparatively simple matter. Since this species ordinarily grows in even-aged stands and requires almost full sunlight to grow well, clear cutting with due regard for the seed supply is the method usually followed. Clear cutting is done on areas of not less than an acre, or very heavy selection cutting is practiced leaving little more than seed trees. Typical

* See *West Coast Tree Farms; The Next Step in Timber Growing in the Douglas Fir Region,* published by the Joint Committee on Forest Conservation, West Coast Lumbermen's Association, and Pacific Northwest Loggers Association, Seattle, 1943.

selective logging is not good Douglas fir forestry, since it is often found in mixture with more tolerant hemlock, spruce, and fir, in which case a selective cut would result in these species making up the bulk of the next crop.

In computing the cost of growing Douglas fir, protection against fire is paramount. Hence an initial investment of $1.00 per acre in improvements and equipment, and a current yearly outlay of 15¢ per acre must be incurred which should be sufficient to hold the annual loss down to one-fourth of one per cent. Then there will be an annual outlay of 12¢ per acre for land rent, taxes, and administration. Oregon and Washington have put into effect a tax system on growing forests which makes the annual charge low and places the main tax on the timber when cut by a yield tax of 12½ per cent of the stumpage value. Planting where necessary is figured at $9.00, and this materially raises the final cost. All costs are carried forward with the interest rate at 3 per cent, compounded annually. Gross yields have been discounted for all risks such as fire, incomplete restocking, etc.

Using the above costs and assuming a yield for Douglas fir on a rotation of eighty years of 74,800 board feet for Site I, 59,900 board feet for Site II, and 41,300 board feet for Site III, the total cost per thousand board feet was computed for three different cases. Case one would start with land satisfactorily restocking with twenty-year-old natural reproduction, and the total cost per thousand feet would be 67 cents, 83 cents, and $1.21 for the three sites respectively. Case two would start with land satisfactorily stocked with one-year-old natural reproduction, and the total cost per thousand feet would be $1.30, $1.63, and $2.35 respectively. Case three would start with bare land and planting 500 trees per acre, in which case the costs would be $2.58, $3.22, and $4.67 respectively.

Fortunately planting is required in only limited areas, and if adequate seed trees and fire protection are provided, could be eliminated from the computations on many tree farms. It is interesting to note that privately owned timberlands in the Pacific northwest will average about Site II in growing quality. To indicate the profits involved in the above operations, it is to be noted that the average stumpage prices for old growth, all species, was $5.75 per thousand board feet in 1942.

Regarding the future use of wood, the industry is trying to combat two false notions quite prevalent among the public at large. There is no foundation for the rumor that wood will be replaced by other materials. Quite the contrary is true. More products are being made from wood now than ever before. Nor is there any reason for believing that there is not enough wood to go around and that we are headed for a timber famine. Wood has played a big part in making America great and will continue to do so. We shall always have enough wood for our needs if present forestry methods are continued and extended, and if both public and government take a sincere, coöperative interest in the forest problem, which affects each and every one of us. Both on this continent and abroad, wood is now regarded as a universal raw material capable of furnishing most of the necessities of life—food, shelter, clothing, fuel, and energy—to say nothing of the other services that forests render, already adequately described.

At the moment war uses are most important. Some day, no doubt, the full story of the part that the lumber industry played in the present world conflict will be written. It will be an impressive narrative. In 1942 the Army, the Navy, shipbuilding, overseas bases, defense housing required over four billion feet of lumber. Wood for bombers and gliders, for army barracks, for hangars, for supply depots, for air fields, for mosquito boats, and for gunpowder are only a few of the general uses of this all-important raw material. And the Douglas fir region furnished a very large part of the total.

But, while lumber has gone to war, forest industries have an eye focused on the future. A number of infant wood-using industries promise to mature into giants after the war is won. Wood will be used as wood in a thousand new ways, but it will also be used in endless variety as cellulose fiber, as sources of chemicals, and as plastic materials. The structural use of wood will continue in heavy volume. New homes must be created by the millions. Plywood glued with synthetic resins will enable us to build better, more quickly, and more economically. Wood and glue have also combined to make boards and dimension stock out of scraps, and 100-foot curved beams and trusses out of one-inch stock. The chemical treatment of wood to resist fire, decay, insects, and unfavorable climatic conditions has just begun. Wood for artificial wool, for industrial alcohol (for synthetic rubber), and for thousands of chemical products such as camphor, charcoal, and lignin products, has great possibilities. The recovery of industrial alcohol, lignin, and other products from pulp-mill waste liquor; wood gas for driving motor vehicles; sugars and protein feeds for cattle are only a few of the new unlimited possibilities in this field.

Hence the outlook for the uses of wood after the war is limited only by the amount of research that can be put into the problem.

No raw material has wider application in the large basic industries. Furthermore, the lumber industry can look forward to a more complete utilization of raw product than has ever been known before, including the use of short-length scraps, small-diameter logs, and even sawdust, trimmings, edgings, and shavings in the fiber, chemical, and plastic industries. This close utilization will greatly reduce the amount of waste, will greatly increase the returns from second-growth forests, and may conceivably reduce the cost of wood products to the ultimate consumer. These new industries will also make possible the utilization of so-called "inferior" species which in the past have often been left in the woods, or at best only incompletely harvested.

In short, we can grow all the wood we need for all uses only if the public will learn one of the elemental lessons of conservation—stop burning up the trees that produce wood. We shall have not only a sustained yield of wood products but also a sustained yield of prosperous human beings. The logging industries and the public must coöperate to produce permanent communities of healthy, happy, busy workers with all the social, cultural, and recreational benefits they have a right to expect from a great democracy.

INDEX

www.ingramcontent.com/pod-product-compliance
Lightning Source LLC
LaVergne TN
LVHW090300210926
848480LV00001B/84